Basic Python In Finance

Bob Mather

Data Analysis from Scratch Python Bundle

2 Books in 1

Bob Mather

1

FREE GIFT

If you're interested in learning how artificial intelligence will take over your world in the next decade, or you're just trying to get automation tools for your office and business, you will love my book titled 'Artificial Intelligence for Business' below:

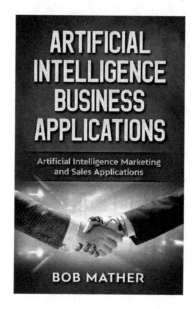

https://bobmatheraibusiness.gr8.com/

4

Table of Contents

Why Python In Finance?

We live in a world where technology has taken root in everything; by everything, we mean literally *everything*. From waking up and to look at your phone to making coffee using your coffee machine, the last decade has seen more inventions and advancements in technology than any other era in human history. Researchers estimate that by 2030, the world will have flying cars to reduce congestion. Amazing, right? And in the excitement of new automation, we have not left business and finance behind, as companies have been turning to technology to stay ahead of their competitors.

In the world of finance, technology is an asset. Technology significantly improves time and efficiency of a business. Companies no longer depend entirely on the financial aspects but look toward new innovations. Technology not only brings out modernization, but it also speeds up the rate of financial transactions and gives out large volumes of data. It will not be wrong to say that technology has become the main distinguisher between institutions.

Programming languages such as R, C+, C++, Java, and Python dominate the game, and this book will cover about everything you need to know about Python in finance. It will cover the following topics among others:

- An introduction to Python and the basics that you will require to get started. Get to know about how to use Python in finance and the benefits that it brings to the table.

- You will learn about the basic trading strategies and time series. Get to know about stock and bonds.

- Time series data and common financial analyses that you will encounter such as volatility calculations, cumulative daily rate of return, moving windows and the dividend calculations among others.

- The common trading strategies that are involved in Python.

This book is mainly aimed at financial professionals and stock investors who wish to get started implementing Python code to automate their finances. It touches on the different areas and their specific codes. Just like Python is an asset to finance, this book is also an asset to you.

Why Python in Finance?

Before we kick everything off and get to the complex parts, we will first begin with examining why Python in finance is important. One reason that makes Python a popular programming language is because it is simple to write, thus making it an excellent tool for traders, analysts, and researchers. A 2018 report by HackerRank 2018 Developer Skills Report showed that the number of financial institutions that were using Python had tripled in the previous two years, i.e. 2016 and 2017. The following are the reasons entrepreneurs are turning to Python for their financial needs.

It is flexible and simple

As mentioned above, Python is easy to understand and deploy. This makes it a perfect tool for handling and dealing with complex financial applications. It is highly accurate and thus reduces the rate of error; a very critical factor in finance, especially when dealing with highly regulated industries. Python is fast, which is a bonus because organizations can build on their software quickly and bring them to the market in no time.

It is rich in tools and libraries

Python has an advantage in that developers need not build tools from scratch. On top of saving the organization tons of cash, it also goes a long way in reducing the time spent on a single development project. If your company's products require integration with third parties, Python has you covered. In short, **it makes everything easier**. Its vast libraries and collections of tools enhance Python's speed, helping to build a competitive structure for organizations seeking to address the needs of consumers that are changing every day by releasing unique products fast enough.

Python is popular

The community behind the development of Python comprises passionate and vibrant developers who have contributed to creating practical tools and organizing many events to share the knowledge and benefits of Python. Each year, the community grows and the number of people opting to use Python is constantly increasing. Experienced developers join the Python community and add to its value, adding to its popularity. Organizations that have invested in Python are certain that technology is here to stay, and there are no signs of it being obsolete anytime in the coming future.

Enables organizations to build their MVPs quickly

Financial organizations need a technology that is scalable and flexible; they need something that will respond quickly to their customers' demands and offer personalized services that will add value to them. Python offers nothing short of this. Python enables developers to create an MVP that will help the organization find a market for their products in no time. The MVP is flawless, and the organization can change or add new codes to suit their products or create new unique ones.

Python bridges data science and economics

Other programming languages like Matlab are less popular with economists who have opted for Python in making their economic calculations. Python's practicality and simplicity in creating formulas and algorithms makes it a prodigy in the game, as Python makes bridging data science and economics simpler than it is.

This is just the tip of the iceberg for why Python is great for finance, and there are still many more reasons you should consider this programming language. One thing for sure is that Python is a giant in the game; moreover, it is simple to write and a great starting point for beginners.

Before going deeper into the various trading strategies, it is essential that we get hold of the basics at first. This book will give you a detailed explanation of Python basics that will go a long way in assisting you to get started. It is essential that you know the basics for Pandas, and have learned your way around NumPy, packages, and Python lists. If you don't know the basics right now, worry not; we've got you sorted.

Chapter Summary

This chapter covered why Python in finance is a great innovation. We went over how financial institutions and data analysts should have a basic knowledge of Python, and we highlighted some other reasons to consider Python.

In the next chapter, you will learn about basic trading strategies and how to invest in various stock assets.

Basic Trading Strategies and Time Series

This chapter will introduce you to various trading strategies using Python. You will also learn business trading using stocks and about time series data.

Financial institutions use Python to make it easier to trade financial **securities** like **bonds**, **stocks**, and tangible products such as gold and oil.

Before we dive into stock trading and time series data, we want to make sure you understand some basics of finance. If you are a well versed financial professional, you may skip this section.

What Are Stocks and Stock Trading?

Stocks

Stocks are equity investments of a particular corporation, and they represent part of ownership of that corporation. When you buy stocks of a corporation, you become a shareholder in that company, which entitles you to receive that corporation's earnings and assets.

The **shareholders** receive profits from the proceeds of the company in the form of **dividends**. If you buy ordinary or common stocks, you will have voting rights at the annual meetings of that company, but there is no guarantee that you will receive dividends. Shareholders who buy **preferred stocks** have no voting rights, but that guarantees they receive dividends from the company proceeds.

In finance, a stock is a **financial security** that represents ownership in terms of shares to a company. The company issues shares at a particular amount; this amount determines your claim on the company's assets and its performance.

Companies raise funds for engaging in their projects through shares to the public. Investors interested in such stocks buy them, thus providing the capital needed for the company.

Investors buy and sell stocks through **brokers** or **online trade**. Before buying any stocks, make sure you evaluate the different stocks to know which is suitable for you. You should also decide how many you need to purchase.

Bonds

When a company raises money through issuing debt or borrowing a loan from a banking institution, they receive *bonds* instead of *stock*. A **bond** is a fixed income instrument representing a loan issued to a **borrower** by the **investor**.

Companies, states, and other sovereign governments use bonds to finance major projects or grow their business. Governments also use bonds when borrowing money. **Debtholders**, or **creditors**, issue these bonds.

Large organizations need more money that average banks can give; therefore, bonds provide a solution to those organizations by allowing several individual investors to assume the role of a **lender**.

Stock Trading

Stock trading involves buying and selling *existing* or *previously issued* stocks. Depending on the company's performance, the share prices can be high or low. Stock prices fluctuate based on the current market price or based on **supply and demand** in the share market.

Traders buy stocks of the company at low prices with anticipation that the prices will rise, allowing them to sell the stocks at higher prices later.

Some traders may use a **short-selling trading strategy**, similar to the concept just described. This strategy is when the trader borrows shares from the company and sells immediately, hoping to buy them later at a lower price, then return the borrowed shares to the lender. In return, traders can realize more profit.

Investment banks and **stock exchange institutions** develop various trading strategies and frameworks to help them make investment decisions. Python is one of the most popular programming languages used to develop and design these trading strategies.

Similar to building a project, developing trading strategies contains a series of steps: from planning the strategy to designing and testing the performance of this strategy.

These trading strategies are used to test historical data based on the rules defined with the strategy before being applied in the real market. Testing will help determine how effective is your strategy and areas you need to improve.

Trading Strategies

The best trading strategy is the **price action strategy**, which will give you the best return on investment. **Forex traders** rely on a variety of trading strategies to determine the best time to invest in stocks and when to sell them. Investors innovate new **trading patterns** and **analytical methods** to make it easier to understand price market movements.

Most investors use a **day trading pattern**, where they buy and sell financial instruments within one day or even have multiple transactions in a single day.

Be careful if you want to use a **day trading strategy**, especially if you are new. The transaction costs and risks with this strategy are very high. Choose the best trading strategy that limits your losses and find a reliable **broker** that will help you trade successfully.

If you rely on **online brokers**, make sure they have an advanced platform that shows real-time streaming of **stock quotes** and **advanced charting tools**. These applications will enable you to modify multiple orders in quick succession.

Basic Principles for Successful Trading strategy

Have the latest stock market news

Knowledge is power, and having knowledge of the latest stock market news and other events will help you succeed. Do a lot of research and inform yourself about the current economic outlook of the country and **interest rate plans**.

Keep a list of the stocks you would like to trade on and the companies to buy these stocks. Read business news from various financial websites to learn the happenings of the general market.

Put aside some funds.

Once you keep a list of stocks to trade, you also have to decide how much you're willing to risk trading. Always set some funds aside which you can use to trade and also prepare yourself for possibly losing them. Trading is a win-lose strategy; you can either win or lose your money.

Set a trading time

Set aside some time for you to use for trading each day. You need time to track the markets and spot opportunities that arise during the **trading hours**. If you're using a day trading strategy, grabbing an open opportunity quickly is the key.

Start small

If you're new in the stock market, starting small is the way to go. You can focus on a maximum of two stocks per go, which makes it easy to track and find opportunities. You can also trade in fractional shares where you can invest in smaller amounts. For example, if a company is trading at $250, you can buy shares worth $50 (one-fifth of a share) from the brokers.

Keep away from penny stocks

Always avoid buying stocks selling at low prices because most of the time these stocks are illiquid and can become delisted from the major stock exchange companies. Most trade these stocks over-the-counter, which means you can easily lose your investment.

Be realistic

When trading, be realistic about the strategy. Not every strategy you win will be profitable; many traders win 50% to 60% of all their trades. Take a limited risk with each trade and define your **entry** and **exit methods** for each trade.

Stick to the plan

To succeed, you need to develop a trading strategy in advance and always stick to it. This will help you move fast and be successful. Your plan will help evaluate your trade closely and avoid chasing profits.

Strategies

Follow the Trend

This means you buy shares when the prices rise and short sell when the price drops. You buy shares with the assumption that the rising and falling of share prices will continue to do so.

Scalping Strategy

In this strategy, a speculator will exploit a small price gap created by the **bid-ask spread**. It allows you to enter and exit the market within a short timeframe.

Contrarian Investment

In the contrarian approach, you buy stocks during their fall in price and short sell during the price rise. Investors operate on the assumption that the rising price will reverse and then drop after a while.

Trending News

This strategy allows investors to buy stocks when there is good news and short sell when there is bad news on the stocks. One has to be careful when using this strategy, as it can lead to both higher profits and losses.

There are pros and cons to all of the above strategies. Do what suits your level of involvement and risk to make it work for you.

Time Series Data

This is a series of data points taken at a particular time interval. In stocks, a time series tracks the changing prices of shares over a specified time period, and it forecasts stock prices over a specified period.

We would usually plot **time series data** via **line charts** and use them in forecasting data trends. Investors carry out a time **series analysis** on time series data to extract meaningful statistics and obtain other characteristics of data. Time series data predicts *future* data values based on the *previously recorded* data values.

The data recorded can be on an annual, quarterly, monthly, weekly, daily, or hourly basis. The investor analyses the data and draws a chart depicting the sequence of observations.

Analysing time series data is a step-by-step preparation process. Beyond analysing future values, it also helps develop a meaningful and accurate forecast of the recorded data.

A **time series chart** is a series of snapshots or events that have taken place at a regular time interval. It is an ideal tool for investors to determine stock trends in the market before investments.

The chart acts as a dashboard or data visualization tool that shows data points at successive time intervals. All data points show the prices and time to measure. The dashboard allows investors to spot trading patterns easily and analyse how the stock market prices are changing.

Importing Time Series Data in Python

We store time series data in the form of a **.csv file**, or a spreadsheet format. The data comprises two columns: the *dates* and the *values*.

We can import this data into a Python program using **Pandas library**. Pandas time series package can easily integrate with any data set and has specific time series data structures for dealing with dates and times. A **timestamp** represents a single point in time. To create a timestamp from strings with a different date/time format, we would use the *to_datetime()* function.

Here is how we would import Pandas and convert a date to a timestamp:

```
import Pandas as pd

pd.to_datatime('2019-09-27 3:45pm')
```

Output

```
Timestamp('2019-09-27 15:45:00')
```

Example 2

```
import Pandas as pd
pd.to_datetime('25/8/2019')
output
Timestamp('2019-25-08 00:00:00')
```

The to_datetime() function tells Pandas to convert to the date/time format based on the input provided. Pandas function automatically converts date/time to **month/day/year**.

pd.to_datetime('25/8/2019') converts to Timestamp('2019-25-08 00:00:00')

Alternatively, you can use the dayfirst= function to tell Pandas to interpret this date as August 25, 2019:

```
pd.to_datetime('25/8/2019, dayfirst=True)


Timestamp('2019-08-25 00:00:00')
```

If you have columns with dates, you can use the Parse_dates=['date'] function to make the data in that field parsed as the date.

Reading Time Series Data in Pandas package

You can import any .csv program into your Python program. Pandas' `read_csv()` method allows you to read a .csv file or any spreadsheet format used. The function imports data into Pandas DataFrame.

Assume you have the following *sales.csv* file; you can import this file into your Python program and manipulate it.

	A	B	C	D
1		Date	Value	
2	0	7/1/1991	3.526591	
3	1	8/1/1991	3.180891	
4	2	9/1/1991	3.252221	
5	3	10/1/1991	3.611003	
6	4	11/1/1991	3.565869	
7				
8				

Open your IDE and type in the following code:

```
from dateutil.parser import parse

import matplotlib as mpl

import matplotlib.pyplot as plt

import seaborn as sns

import numpy as np

import pandas as pd

plt.rcparams.update({'figure.figsize': (10, 7), 'figure.dpi':120})

dframe=pd.read_csv('sales.csv', parse_dates=['date']

dframe. head()
```

	date	value
0	1991-07-01	3.526591
1	1991-08-01	3.180891
2	1991-09-01	3.252221
3	1991-10-01	3.611003
4	1991-11-01	3.565869

Plotting Time Series Data in Python

Time series visualization plays an important role in data analysis and forecasting. Plots of raw data can provide insights into trends, cycles, performance, and relationships between variables. Visualization of data can be in the form of graphs, line plots, histograms, etc.

Using the daily minimum temperature dataset, this is how to use the dataset for time series visualization. You can download a sample file to practise this code by clicking on this <u>link.</u>

If you're reading the paperback, you can obtain this dataset at:

https://raw.githubusercontent.com/jbrownlee/Datasets/master/daily-min-temperatures.csv

```
from pandas import read_csv
from matplotlib import pyplot
#Display dataset for default 5 rows
temp= read_csv('daily-minimum-temperatures.csv',
            header=0, index_col=0, parse_dates=True, squeeze=True)
print(temp.head())
```

When you run this dataset, it will print the **first 5 rows** of the daily-minimum-temperature dataset.

```
Date
1981-01-01 20.7
1981-01-02 17.9
1981-01-03 18.8
1981-01-04 14.6
1981-01-05 15.8
Name: Temp, dtype: float64
```

We can now plot a line graph to represent the data.

```python
from pandas import read_csv

from matplotlib import pyplot

temp = read_csv('daily-minimum-temperatures.csv',

            header=0, index_col=0, parse_dates=True, squeeze=True)

temp.plot()

pyplot.show()
```

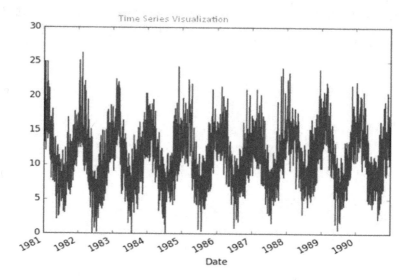

Minimum Daily Temperature Line Plot

Time Series Data Plotted Using a Histogram

```
from pandas import read_csv
from matplotlib import pyplot
temp = read_csv('daily-minimum-temperatures.csv',
        header=0, index_col=0, parse_dates=True, squeeze=True)
temp.hist()
pyplot.show()
```

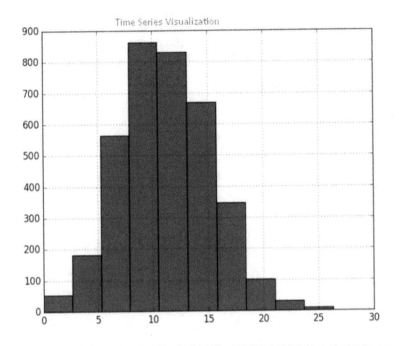

Minimum Daily Temperature Histogram Plot

Chapter Summary

In this chapter, we learned about various trading strategies you can use when investing in the stock market.

- Seven principles to guide you in your daily trading.

- Time series data and how to use it.

- Importing time series data into Python.

In the next chapter, you will learn how to install and set up the Python program into your device.

Setting Up the Workspace

Setting up your workspace does not have to be rocket science and is something you can do entirely on your own. There are requirements that you will need, such as an **Integrated Development Environment (IDE)** and **Python** with you. In this chapter, we will go over the easiest way to set up your workspace for Python.

Using Anaconda is a nice way of starting out. **Anaconda** is a high-performing distribution of both the R and Python programming languages. With it comes hundreds of popular Scala, R, and Python packages you can use for data science. That's not all—there are lots of other benefits accrued with Anaconda too. Installing it will give you access to over 720 other packages for your data science work and any other statistical data. You will also have access to **Jupyter Notebook** and **Spyder IDE**.

Compared to other programming languages, many beginners prefer Python. It has a myriad of libraries with many modules that enable you to analyze your stock data without the need for writing the function code.

Essential Components of Python

Anaconda: Downloading and installing individual libraries and tools needed for your statistical data may be tedious, thus, we need to install Anaconda. Anaconda has all the tools and libraries needed to execute any Python code, with various Python packages to load directly to your IDE framework.

Spyder IDE: Spyder IDE acts as the code editor or software platform where you would write and execute your program codes. The IDE (Integrated Development Environment) not only presents you with the code window to write your code, but it also has a compiler and interpreter used to convert the human-readable code to machine-readable language. It has a debugger program that scans your program for any bugs or errors in the code.

Jupyter Notebook: This is an open-source program that enables you to write and implement program codes in a more interactive format. Jupyter Notebook is used to test a small program code, whereas Spyder IDE is preferable when dealing with multi-file projects.

Conda: Installing Conda will enable you to update your libraries easier. It acts as a management system that automatically installs, runs, and updates your library.

Spyder IDE and Jupyter Notebook will install automatically once you install Anaconda, as they are parts of the Anaconda Distribution.

Installation Process

Step 1

Open the Anaconda website. Click on the version you want to download based on your operating system and system specifications (64-bit and 32-bit).

In this tutorial, we will be downloading Anaconda for **Windows**.

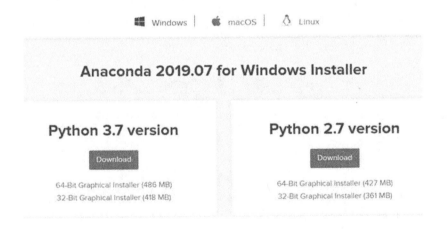

Step 2

Once the download is complete, locate the download file and click it to run it.

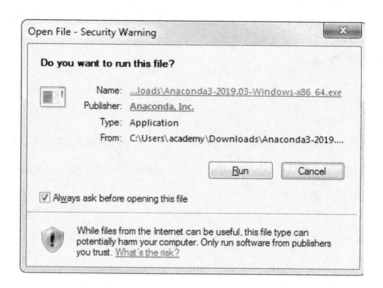

Click **Run** to start the installation.

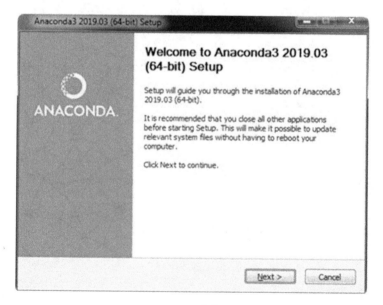

Click **Next** on the opened installation wizard and accept the agreement terms by clicking **I Agree**.

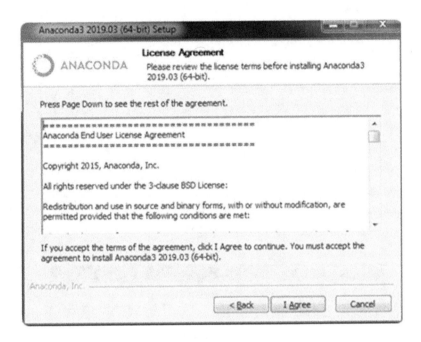

Step 3

Under the Installation type, select **Just Me (recommended)**, then click **Next**. You want to **Save the file**. You can specify where you want to save Anaconda (via **Browse**), then click **Next**.

Step 4

Under **Advanced Options**, select the **Register Anaconda as my default Python 3.7** check box, then click **Install**. Once installation is complete, click **Finish**.

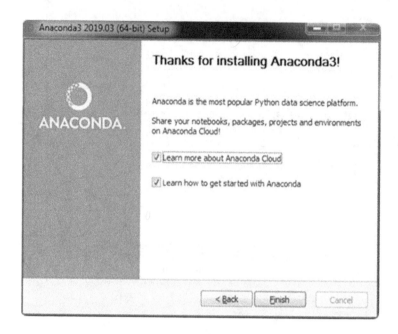

After successfully installing Anaconda, you can open the program navigator and see some of the already installed tools like Jupyter Notebook and Spyder IDE.

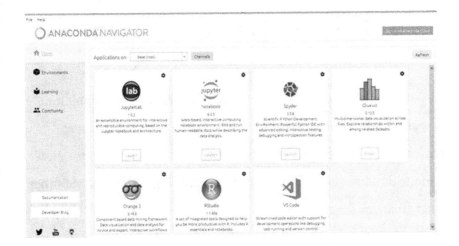

If you're using **MacOS** or **Linux**, you can find installation details on the official Anaconda website.

Chapter Summary

In this chapter, we went over the step-by-step process for installing Anaconda and learned about the various components essential for running the Python program.

Python Basics for Finance

Python for finance allows you to use various data manipulation packages to perform functions such as analyzing statistical data, visualization, and machine learning, among others. In this tutorial, we will focus on libraries required for coding trading strategies and other popular libraries for finance.

A library comprises reusable modules and functions that can perform specific functions. There is no need to write code from scratch for these functions, making Python easy-to-use in data and time series analysis.

Python Packages/Libraries

Pandas: Pandas represent data in the form of a **DataFrame**, with tables or spreadsheet format where you would store data in rows and columns. It allows you to import data directly from an Excel or .csv file into the Python code window. This makes it easy to do data analysis and manipulation of the tabular data.

NumPy: The NumPy library, or **NumericalPy**, is used to perform any numeral calculation on arrays of data. In programming, we call a group of elements an **array**. We can perform different operations using the function NumPy on array elements.

Matplotlib: If you want to plot 2D graphs like line charts, bar charts, histograms, and other charts, we call the *Matplotlib* function. The library has various functions that allow you to customize or modify the graph based on your preferences.

Zipline: We use the Zipline library in trading applications. It is an open-source, event-driven system that allows you to receive live trading patterns and perform backtesting of trading strategies.

Technical Analysis Library: Technical Analysis library, or simply **TA-Lib**, allows you to perform technical analysis on financial data. It uses technical indicators like the **Relative Strength Index (RSI)**, **Bollinger Band**s, and **MACD**, among others for analysis.

These Python libraries are ideal for bank analysts and stockbrokers for perfecting their trading strategies.

How to Work with Data in the Python Program

Knowing how to use Python in retrieving data or formatting data is essential for any financial analysis and trading activity.

Python Pandas allows you to analyze time-series data easily and to import data from an Excel file or CSV file format. Investors commonly refer to trading data as time series data because the file would usually index by time (yearly, monthly, weekly, daily, or by the minute).

To carry out any data manipulation and analysis, we need to:

- Import financial data

- Perform numerical analysis

- Build trading strategies based on time series data values

- Plot graphs (time series graphs)

- Perform backtesting on the data

Importing Financial Data in Python

In the following sections, we will import data from different sources using Pandas package.

How to Import CSV Files

The Pandas library has a lot of CSV parsing capabilities that allow you to carry out any numerical analysis. You don't have to build a **CSV parser** from scratch.

CSV file format is a convenient way to export large volumes of data from a spreadsheet or database. We can import this data to use it in another program. When importing CSV files, ensure `pd.DataFrame()` has the same CSV source file.

Pandas DataFrame makes reading CSV files a quick and straightforward process.

Syntax

```
import pandas as pd


df = pd.read_csv (r' Path to the CSV file stored in your
device/Filename.csv')

print (df)
```

This is a general code for opening any CSV file in Python. You only need to change your CSV file path and run the program.

The `pd.read_csv()` opens, analyzes, and reads the contents of the CSV file name and stores this data into DataFrame.

Example

The table below shows a list of stock prices for four companies stored in a CSV file (with file name as *NSE_listing.csv*).

	A	B	C	D	E
1		Stock Prices per share			
2	Company	2017 Prices	2018 Price	2019 Prices	
3	Safaricom PLC	$25	$28	$30	
4	Telcom Kenya	$18	$20	$19	
5	Mumias Sugar ltd A	$15	$20	$12	
6	Kenya orchards ltd	$20	$26	$38	
7	Flame tree group holdings	$15	$10	$17	
8	Carbacid Investments ltd	$28	$20	$24	
9	East Africa Breweries ltd	$19	$25	$24	
10	Total	$140	$149	$164	
11					

The following steps outline how to import this file using Python Pandas.

Step 1: Capture the CSV File Path

This is the path where the file is stored. In this case, the file is stored under the following path:

C:/Users/Faith/Documents/NSE_listing.csv

Make sure the file name written in the code matches the file name of your stored file and that it has the **.csv file extension**. This is essential when importing CSV files.

Step 2: Write the Python Code

Open your Python IDE framework to type the code.

```
import pandas as pd

df=pd.read_csv(r'c:\Users\Faith\Documents\NSE_li
sting.csv')

print (df)
```

Step 3: Run Your Code

When you run the code, you will get something like this:

Company	2017 Prices	2018 Prices	2019 Prices
Safaricom PLC	$25.00	$28.00	$30.00
Telcom Kenya	$18.00	$20.00	$19.00
Mumias Sugar ltd A	$15.00	$20.00	$12.00
Kenya orchards ltd	$20.00	$26.00	$38.00
Flame tree group holdings	$15.00	$10.00	$17.00
Carbacid Investments ltd	$28.00	$20.00	$24.00
East Africa Breweries ltd	$19.00	$25.00	$24.00
Total	$140.00	$149.00	$164.00

Customizing the CSV File

You can customize your CSV file and choose only specific rows or columns that interest you.

Let's write a code that will only display the company and 2018 stock prices from NSE_listing.csv file.

```
import pandas as pdr

df=pdr.read_csv(r'c:\Users\Faith\Documents\NSE_l
isting.csv')

df=pdr.DataFrame(data,columns=['company', '2018
prices'])

print (df)
```

Company	2018 Prices
Safaricom PLC	$28.00
Telcom Kenya	$20.00
Mumias Sugar ltd A	$20.00
Kenya orchards ltd	$26.00
Flame tree group holdings	$10.00
Carbacid Investments ltd	$20.00
East Africa Breweries ltd	$25.00
Total	$149.00

Once you import the data into Python, you can start statistical calculations using Pandas.

Data Manipulation Using Time Series Data

Data manipulation is a technique used in organizing data to make it easier to read. You can easily manipulate large volumes of data using the Pandas library. In this example, we will use time series data to analyze data and present it in an orderly manner.

We will use **open system data (OPSD)** on *power consumption* in *Germany*. This dataset has a countrywide power consumption for the period between 2007 and 2017. To follow up with the example, download the dataset here.

Let's start by reading the OPSD file:

```
#read opsd file

opsd_daily = pd.read_csv('opsd_germany_daily.csv')

opsd_daily.head(3)#to return the first three rows

opsd_daily.shape #displaying power consumption for three units
```

	Date	Consumption	Wind	Solar	Wind+Solar
0	2006-01-01	1069.184	NaN	NaN	NaN
1	2006-01-02	1380.521	NaN	NaN	NaN
2	2006-01-03	1442.533	NaN	NaN	NaN

When you run this code, it will return the number of rows in the dataset. The *head()* function tells the dataframe to display the first three rows in the dataset.

Customization of Time Series Data

To customize the data, we will import **Matplotlib** for plotting data, and **Seaborn library**, which will provide us with unique ways for styling and adding appropriate size to the plots.

```python
import matplotlib.pyplot as plt

import seaborn as sns

# Use seaborn library for defaults styles and set the default
figure size of the plot

sns.set(rc={'figure.figsize':(10, 4)})

opsd_daily['Consumption'].plot(linewidth=0.5);
```

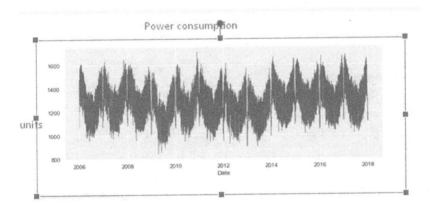

From this plot, you can see that the line plot is too crowded because there are too many data points, making it hard to read. We can solve this by plotting the data in the form of dots and separating the power consumption into *solar* and *wind*.

```
import matplotlib.pyplot as plt

import seaborn as sns


#power consumption for solar and wind

cols_plot = ['Consumption', 'Solar', 'Wind']

#change plot color and set figuresize

axes = opsd_daily[cols_plot].plot(marker='.', alpha=0.5,
linestyle='None', figsize=(10, 9), subplots=True)

#plotting daily power consumption

for ax in axes:

    ax.set_ylabel('Daily Totals (GWh)')
```

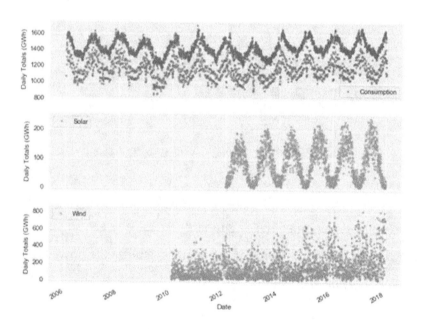

From the plot, we can see that there is high consumption in winter because of increased heating. Wind energy production is also high in winter and low in summer, and so is electricity. Solar production is also high during summer periods.

Importing Google Finance

You can also use Python to download financial data from Google sources or free online databases. Once we download the data, we can use the Python Pandas package for data manipulation and manipulate the data. We then create some data indicators that we can use for the basis of quantitative analysis.

Once you install Anaconda, it will automatically install Pandas and Matplotlib packages. If for some reason the Python distribution does not include them, you can use the *pip* and *conda* packages to install. If you already have the Pandas package installed, you only need to import it into the new project you're working on.

If you're using a non-Anaconda program, you can use *Pandas_datareader (pip install Pandas-datareader)* package and *matplotlib* package for data visualization.

To test whether the *pip install Pandas* command works, open the *'cmd.exe'* command prompt and type: *pip install Pandas.*

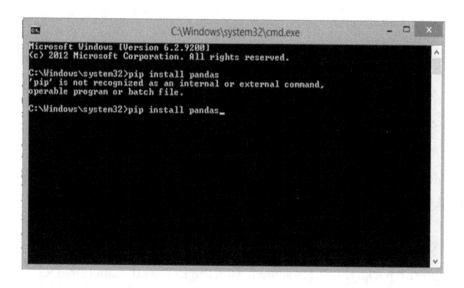

If you're using Windows 64-bit, you will get an error message that *'pip' is not recognized as an internal or external command.* Don't worry, as this only means that the *pip* command is not on your path.

You can change the command path to *C:/User/Abi/Python34/Scripts/pip install Pandas* and run the command. *Python34* in the command path means you're using Python 3.4. If you're using version 3.7 and so on, the code will change to *Python37*.

If you installed your Python program via Anaconda, you can install all the additional models using `conda install <module name>`. Add the program path to your file and replace the module name with the actual name of the module you want to install.

In this tutorial, I have installed Anaconda, so to install a module, I will use the *conda* install command.

C:/Users/Abi/Anaconda3/Scripts/conda install Pandas

```
                    C:\Windows\System32\cmd.exe              _ □ ✕

operable program or batch file.

C:\Windows\System32>C:/users/faith/Anaconda3/Scripts/conda install pandas
Traceback (most recent call last):
    File "C:\users\faith\Anaconda3\lib\site-packages\conda\exceptions.py", line 10
62, in __call__
        return func(*args, **kwargs)
    File "C:\users\faith\Anaconda3\lib\site-packages\conda\cli\main.py", line 84,
in _main
        exit_code = do_call(args, p)
    File "C:\users\faith\Anaconda3\lib\site-packages\conda\cli\conda_argparse.py",
line 80, in do_call
        module = import_module(relative_mod, __name__.rsplit('.', 1)[0])
    File "C:\users\faith\Anaconda3\lib\importlib\__init__.py", line 127, in import
_module
        return _bootstrap._gcd_import(name[level:], package, level)
    File "<frozen importlib._bootstrap>", line 1006, in _gcd_import
    File "<frozen importlib._bootstrap>", line 983, in _find_and_load
    File "<frozen importlib._bootstrap>", line 967, in _find_and_load_unlocked
    File "<frozen importlib._bootstrap>", line 677, in _load_unlocked
    File "<frozen importlib._bootstrap_external>", line 728, in exec_module
    File "<frozen importlib._bootstrap>", line 219, in _call_with_frames_removed
    File "C:\users\faith\Anaconda3\lib\site-packages\conda\cli\main_install.py", l
ine 8, in <module>
        from .install import install
```

We can now write a code to import data:

```
import Pandas as pd

import datetime

import matplotlib.pyplot as plt

import Pandas_datareader.data as web
```

The `Import Pandas as pd` command imports the Pandas module. *Import datetime* is for when we want to retrieve data based on specific dates and tells Pandas module which dates data it needs to generate.

If you pulled the data from the Internet, use the web command `import Pandas.io.data`.

Once you import the data, you can find data on any free online source such as Google Finance or Yahoo Finance.

In this tutorial, we will use **Google Finance** to obtain market data.

To import Google Finance, you need to install the *googlefinance.client* library. *googlefinance.client* is a Python client library for *google finance api*.

Installing the Library

```
pip install googlefinance.client
or
$ conda install googlefinance.client


#import the libraries
import Pandas as pd
import datetime


import Pandas_datareader.data as web  #to import data from
internet

#Extract price data and time-series data from googlefinance.client

from googlefinance.client import get_price_data, get_prices_data,
get_prices_time_data

# create stock data variables

param = {

    'q': ".DJI", # represent Stock symbol (ex: "AAPL")

    'i': "86400", # Show time Interval size in seconds ("86400" =
1 day intervals)

    'x': "INDEXDJX", # Show type of traded Stock (example: "NASD")

    'p': "1Y" # represent Period in which stocks are traded(Ex:
"1Y" = 1 year)

}

# get price data (return data stored in Pandas dataframe)

df = get_price_data(param)

print(df)
```

Output

```
                         Open      High       Low     Close     Volume
2016-05-17 05:00:00   17531.76   17755.8  17531.76  17710.71   88436105
2016-05-18 05:00:00   17701.46  17701.46  17469.92  17529.98  103253947
2016-05-19 05:00:00   17501.28  17636.22  17418.21  17526.62   79038923
2016-05-20 05:00:00   17514.16  17514.16  17331.07   17435.4   95531058
2016-05-21 05:00:00   17437.32  17571.75  17437.32  17500.94  111992332
...                       ...       ...       ...       ...        ...
```

Example 2

Here, we will use Google Finance to obtain market data. We will use **ETF "SPY"** as the proxy for **S&P 500**.

```python
# import library

import Pandas as pd

import datetime

import Pandas_datareader.data as web

from googlefinance.client import get_price_data, get_prices_data,
get_prices_time_data


# params = [
    # obtaining market data
    {'q': ".DJI", 'x': "INDEXDJX",},
    # Type of stock traded
    {'q': "NYA", 'x': "INDEXNYSEGIS",},
    # S&P 500
    {'q': ".INX", 'x': "INDEXSP",}]
period = "1Year"
# extract stock volume dataframe [open, high, low, and close,
volume data]
df = get_prices_data(params, period)
print(df)
```

	.DJI_Open	.DJI_High	.DJI_Low	.DJI_Close	.DJI_Volume
2016-07-20	18503.12	18562.53	18495.11	18559.01	85840786
2016-07-21	18582.70	18622.01	18555.65	18595.03	93233337
2016-07-22	18589.96	18590.44	18469.67	18517.23	86803016
2016-07-23	18524.15	18571.30	18491.59	18570.85	87706622
2016-07-26	18554.49	18555.69	18452.62	18493.06	76807470
...	...	...	...	...	...

Use of Excel Spreadsheet in Data Manipulation

Excel is a great tool for data manipulation and analysis. If you do a lot of data cleaning and reformatting, you may find integration of excel Python scripting language is of great use. But there is a lot of data manipulation you can do using the simple excel spreadsheet.

You can use various excel features to manipulate data like filtering an excel report into specific fields of interest or use excel function to perform statistical and mathematical calculations.

A basic excel manipulation range from inserting and deleting columns in an excel spreadsheet or just moving data from one column to another. Whereas in advanced data manipulation, you can create several worksheets, copy pasting data from one worksheet to another or using excel functions to manipulate data from worksheet 1 in worksheet 2.

Excel terminologies

Workbook: This is an excel file that houses all data created in the worksheets. It enables you to manipulate the data entered in the sheets or do some calculations. Once you open an excel app, it automatically opens a workbook you can work on and allow you to store your data.

Worksheet: A worksheet is a sheet within a workbook. You can have more than one worksheet in a workbook. Whenever you open a new worksheet, it creates a tab at the bottom of the screen. The tab also shows the active worksheet you're working on. You can also rename the sheet to make it easy to identify each worksheet.

Rows and columns: They show how data is aligned in the cells. In a row, cells are aligned horizontally while a column is aligned vertically.

Cell: It is a rectangular box that used to enter data. A cell consists of numerical values or text data. An active cell is one that is opened for editing and is referenced by its row and column number. For example, cell A2 points to the first row and second column of your sheet.

Workspace: Just like in worksheet where you can open worksheets, a workspace enables you to work on multiple files simultaneously.

Cell range: This is a group of cells in a worksheet which are selected based on a certain criteria. A colon is put between the selected cell references. For example, A2:C4 is a range of cells that tells excel formula to look for data between the cells A2 and C4.

Manipulating Rows and Columns

When you want to manipulate any data in excel, you have to select a range of cells with data that you want to manipulate. Once you select a row or column, you can cut/copy and paste data in those cells. You can also insert new rows and columns to the left or right of the selected row/column.

You can delete a row or column from the worksheet. Deleting a row or column removes that row or column entirely and collapse the entire sheet making the remaining rows/column take up space for the deleted row /column.

Hide and unhide rows and columns. If you have a large spreadsheet you're working on, you can temporarily hide and unhide rows/columns not needed at the moment. This makes it easy to manage the spreadsheet as you work on it.

Format cells within a row/column by either changing text color, font, background colors or text alignment.

Excel menu offers you multiple tools you can use to manipulate data like applying conditional formatting to your data, create pivot reports or use charts for data visualization and use of excel formulas for mathematical calculations or handling statistical data.

Another way to manipulate excel data is through the use of filters.When filtering data, you have to select the rows/columns you want to filter then from the home menu tool you can filter selected rows/column based on a certain criteria.

For advanced data manipulation, you can integrate excel with Python and perform further manipulation.

How to Integrate Python Data with Excel

Microsoft Excel is a powerful spreadsheet application that allows you to produce and work with lots of mathematical and statistical datasets. When working with large volumes of data, your tasks in Excel may become repetitive, which can consume a lot of your time.

You can rectify this problem by automating your Excel task. You can integrate an Excel spreadsheet with a Python program to modify your files and avoid repetitive work. The **openpyxl module** allows you to read and modify your Excel files.

Openpyxl library also makes it easy to call external functions from Excel like COM objects. In using openpyxl, you don't have to go through a thousand rows to find specific ones based on your criteria. Python eliminates the boring and repetitive tasks.

Installing Openpyxl Module

You can view Python documentation on how to install third-party modules into the Python program. To install the openpyxl module, you would simply type `conda install openpyxl` in your IDE. The command would look like this:

```
C:/Users/Faith/Anaconda3/Scripts/conda install openpyxl
```

Once you install the module, you can test whether it is installed properly by typing the following command in your code editor or IDE:

```
>>>import openpyxl
```

If you have succeeded in installing the module, there will be no error message returned when you type the command on the interactive shell.

Creating a Simple Excel Worksheet

```
from open openpyxl import Workbook
workbook = Workbook()
sheet = workbook.active


#insert data into your sheet 1 under cell A1 and B1
sheet["A1"] = "hello"
sheet["B1"] = "world!"


#display the entered data
workbook.save(filename="hello_world.xlsx")
```

This code will create an Excel sheet with the file name "Hello_world.xlsx".

If you open the file in Excel, this is what you will see:

Reading Excel Documents with Openpyxl

Assume that you have an Excel file called "**fruits.xlsx**" with the following data:

	A	B	C	D	E	F	G
1	Fruits	Quantity	Price	Purchase Date			
2	Apples	73	20	4/8/2019			
3	Pears	14	30	5/2/2019			
4	Cherries	85	40	6/5/2019			
5	Oranges	52	10	20/4/2019			
6	Apples	152	20	10/6/2019			
7	Bananas	23	15	8/3/2019			
8	Strawberr	98	35	10/4/2019			
9							
10							
11							

You can open the above file using Python openpyxl module. Type the code below to help open fruits.xlsx by calling *openpyxl.load_workbook()* function.

```
import openpyxl

wb = openpyxl.load_workbook('fruits.xlsx') #opening excel file on a
workbook wb.

type(wb)

<class 'openpyxl.workbook.workbook.Workbook'>
```

The function *openpyxl.load_workbook()* opens the filename and returns a value of the workbook data type. You can also use *workbook.sheetnames* to display the available worksheets.

After opening the spreadsheet, you can easily retrieve specific data from a cell or row. If you need the data in **cell A4**, you can easily retrieve it like this:

```
# retrieve data from cells
sheet.cell(row=4, column=2)

<Cell 'Sheet 1'.A4>


sheet.cell(row=4, column=2).value
```

You can also loop the function to print a series of values in a specific range:

```
# retreive data from cell A2
sheet.cell(row2, column=1)
<Cell Sheet1.A2>
sheet.cell(row=2, column=1).value

#printing data in cells in the range of 1,8,2
 for i in range(1, 8, 2):
     print(i, sheet.cell(row=i, column=1).value)
```

Output

```
output

1 Apples
3 Pears
5 Apples
7 Strawberries
```

Calling the `cell()` method and passing it to `row=2`, `column=1` will display the object in **cell A2**. You would use the `for` loop to print a series of cells based on specific criteria.

Always make sure you supply the exact path to the filename in your code; otherwise, it will generate an error code if the file path is different from the one you wrote in the code.

Dealing with Different Worksheets

If your workbook has data from different worksheets, you can receive data from each worksheet by using the `get_sheet_names()` function.

Example

```
import openpyxl
wb = openpyxl.load_workbook('fruits.xlsx')
wb.get_sheet_names()
['Sheet1', 'Sheet2', 'Sheet3']
#specify the sheet name to get data
sheet = wb.get_sheet_by_name('Sheet3')
sheet <Worksheet "Sheet3">
type(sheet) <class 'openpyxl.worksheet.worksheet.Worksheet'>
sheet.title
# set up active worksheet
anotherSheet = wb.active
anotherSheet
<Worksheet "Sheet1">
```

You can call each worksheet by passing the worksheet name string to `get_sheet_names()`.

You can set any of the worksheet to be at the top anytime you open your workbook by using `.active` feature.

Importing Data from Excel File into Pandas

You can read data from an Excel file using Pandas module. To do this, you need to first import data from your Excel file into Pandas module:

```
Import Pandas as pd
```

After importing, you can then use the `read_Excel()` function to read the data from the Excel file. You can do this by calling the Excel file name using the following command.

```
Excel_file='fruits.xlsx'
Fruits=pd.read_Excel(Excel_file)
```

The call to function `read_Excel()` reads the Excel data into the Pandas DataFrame object since Pandas stores data in form of

DataFrames. After reading the file, Pandas then stores the DataFrame into the variable `fruits`.

Chapter Summary

In this chapter, we learned how to use Pandas to analyze time series data. We can now also import, export, and manipulate data using Pandas.

You can use **Pandas-reader** to read data files in Google, Yahoo Finance Report, and World Bank.

In the next chapter, we go over how to use time series to manipulate data.

Working With Time Series Data

From the previous chapter, we learned how to import data into our workspace. In this chapter, we will learn how to manipulate data using the Python packages Pandas, NumPy, SciPy, and Matplotlib, among others.

By the end, you will also be able to perform some financial analysis from the data imported into your workspace.

Using Pandas-datareader to Import Data

Over the years, Python has been a great language for data analysis and visualization. It's fantastic ecosystem of packages has been a great help in performing statistical and mathematical functions among other functions. Pandas is one of the most commonly used packages for importing and analyzing data in Python.

The Pandas-datareader function allows you to read data from a variety of sources such as Excel data, CSV files, and web data from sources like Google, Yahoo, and World Bank. The Pandas package will allow you to analyze the CSV files and web data imported to your workspace.

Installing Pandas

To start, install Pandas package into your machine using the *pip* command if using Python, or *conda* if using Anaconda.

```
pip install Pandas
```

or

```
conda install Pandas
```

To read Python data, install the *Pandas-datareader* function:

```
import Pandas_datareader as pdr
```

Example

Let's use *Pandas-datareader* to pull data from a Yahoo page:

```
import Pandas_datareader as pd

import datetime

aapl = pd.get_yahoo_data('AAPL',

                    start_date=datetime.datetime(2011, 10,
1),

                    end_date=datetime.datetime(2019, 10,
1))
```

You can also use **Quandl library** to import data from a web source, such as Google Finance data:

```
import quandl

import datetime

aapl = quandl.get("WIKI/AAPL", start_date="2011-10-01",
end_date="2019-10-01")
```

As we have seen, you can use the *Pandas_datareader* command to extract data from any source into your Python program

The program stores the extracted data in a DataFrame *aapl*, which is a data structure that holds both the rows and columns of your data.

Creating Pandas DataFrame

A DataFrame is a tool used in the representation of data in the form of a table. **Tabular data** is data shown in rows and columns. When importing data into Pandas, you need to load the CSV files into Pandas DataFrame.

To create a DataFrame, you need to pass multiple series of data using the $pd.series$ method into DataFrame class. This is like a two-dimensional array.

Example

Here, we will use two series objects: **L1** and **L2**.

```
import pandas as pd

# series variables

L1= pd.Series([1,2])

L2= pd.Series(["Mary", "John"])



#frame the series objects

df= pd.DataFrame([L1,L2])
```

df

	0	1
0	1	2
1	Mary	John

```
# create dataset with use of index and column
values
```

```python
df = pd.DataFrame([[1,2],["Mary", "John"]],

        index=["r3", "r4"],

        columns=["c3", "c4"])
```

df

	C3	C4
r3	1	2
r4	Mary	John

```python
#create dataset with dict-like container

df = pd.DataFrame({

        "c3": [1, "Mary"],
```

```
        "c4": [2, "John"]})
```

df

	C3	C4
0	1	2
1	Mary	John

After creating a DataFrame for your project, you can run the `head()` and `tail()` functions to enable you call the first row and last row of the data stored in the DataFrame.

Example: Importing data from a CSV file to your computer

Assuming you have a file named project_data.csv in your computer, you can easily retrieve it using pandas dataframe. Make sure to use the specific path to the file location.

From the dataframe we can create a code to only retrieve the first 10 rows of data. The head () function returns the first five rows by default. If you need more rows you have to specify the number.

```
import pandas as pd
```

```
# read the dataframe and assign into variable df
dframe = pd.read_csv("Project_data.csv")
```

```
dframe.head(10)
```

Project Name	Start Date	End Date	Duration	Team Members	Project Budget	Actual Budget
Project A	5/5/2016	7/1/2016	57	10	$1,000,000.00	$880,000.00
Project B	5/10/2016	8/10/2016	92	2	$900,000.00	$920,000.00
Project C	6/10/2016	3/1/2017	264	4	$860,000.00	$880,000.00
Project D	6/22/2016	8/4/2016	43	5	$1,000,000.00	$998,050.00
Project E	7/14/2016	11/1/2016	110	7	$294,000.00	$280,000.00
Project F	7/14/2016	1/20/2017	190	5	$123,400.00	$125,000.00
Project G	8/1/2016	10/1/2016	61	12	$250,500.00	$236,000.00
Project H	8/14/2016	8/30/2016	16	2	$127,200.00	$126,000.00
Project I	9/1/2016	12/10/2016	100	7	$80,000.00	$79,900.00
Project J	10/1/2016	11/15/2016	45	5	$77,000.00	$77,000.00

To print the entire rows and columns in the dataframe we use
`dframe.shape` command

```
import pandas as pd

dframe=pd.read_csv("project_data.csv")

dframe.shape
```

Project Name	Start Date	End Date	Duration	Team Members	Project Budget	Actual Budget
Project A	5/5/2016	7/1/2016	57	10	$1,000,000.00	$880,000.00
Project B	5/10/2016	8/10/2016	92	2	$900,000.00	$920,000.00
Project C	6/10/2016	3/1/2017	264	4	$860,000.00	$880,000.00
Project D	6/22/2016	8/4/2016	43	5	$1,000,000.00	$998,050.00
Project E	7/14/2016	11/1/2016	110	7	$294,000.00	$280,000.00
Project F	7/14/2016	1/20/2017	190	5	$123,400.00	$125,000.00
Project G	8/1/2016	10/1/2016	61	12	$250,500.00	$236,000.00
Project H	8/14/2016	8/30/2016	16	2	$127,200.00	$126,000.00
Project I	9/1/2016	12/10/2016	100	7	$80,000.00	$79,900.00
Project J	10/1/2016	11/15/2016	45	5	$77,000.00	$77,000.00
Project K	10/1/2016	12/1/2016	61	10	$65,000.00	$55,000.00
Project L	11/1/2016	12/1/2016	30	4	$550,000.00	$651,000.00
Project M	11/10/2016	12/10/2016	30	3	$45,000.00	$42,000.00
Project N	12/1/2016	2/10/2017	71	1	$32,500.00	$25,000.00
					$5,404,600.00	$5,374,950.00

Indexing DataFrames in Pandas

Indexing means to extract specific rows and columns of data from the DataFrame. This allows it to customize your file by only reading the specific data needed. In indexing, you can select a specific row and column, or select all the rows and a few columns.

Sometimes we would refer to indexing as a **subset selection**, as you would only select a specific number of rows and columns in the process.

Example

Let's use our ***project_data.csv*** DataFrame to practice on how to do indexing. Create the file and follow this example to perform indexing.

	Project Name	Start Date	End Date	Duration	# of Team Members	Project Budget	Actual Budget
1							
2	Project A	05/05/16	07/01/16	57	10	$1,000,000	$880,000
3	Project B	05/10/16	08/10/16	92	2	$900,000	$920,000
4	Project C	06/10/16	03/01/17	264	4	$860,000	$880,000
5	Project D	06/22/16	08/04/16	43	5	$1,000,000	$998,050
6	Project E	07/14/16	11/01/16	110	7	$294,000	$280,000
7	Project F	07/14/16	01/20/17	190	5	$123,400	$125,000
8	Project G	08/01/16	10/01/16	61	12	$250,500	$236,000
9	Project H	08/14/16	08/30/16	16	2	$127,200	$126,000
10	Project I	09/01/16	12/10/16	100	7	$80,000	$79,900
11	Project J	10/01/16	11/15/16	45	5	$77,000	$77,000
12	Project K	10/01/16	12/01/16	61	10	$65,000	$55,000
13	Project L	11/01/16	12/01/16	30	4	$550,000	$651,000
14	Project M	11/10/16	12/10/16	30	3	$45,000	$42,000
15	Project N	12/01/16	02/10/17	71	1	$32,500	$25,000
16						$5,404,600	$5,374,950

Assume that you only want to see data from **Project A**, **Project B**, **Project C** and the corresponding three columns namely; **project name**, **start date** and **end date**. You can write a code that will only retrieve the specific data you need from the dataframe.

Project Name	Start Date	End Date I
Project A	5/5/2016	7/1/2016
Project B	5/10/2016	8/10/2016
Project C	6/10/2016	3/1/2017

There are different methods you can use to pull data from a DataFrame. Python Pandas library offers three ways of indexing data, with the most common being the `.loc()` and `.iloc()` methods. These methods allow you to retrieve as many rows and columns as you need by their position.

`DataFrame. []`: This method acts as the **indexing operator**.

`DataFrame.loc[]`: Use the `.loc[]` function when dealing with **labels**.

DataFrame.iloc[]: The `.iloc[]` method is great for when you're looking for the position of certain DataFrames or when dealing with **integer-based data**.

How to use DataFrame indexing operator *[]*

We designate an operator as the **indexing operator** by following it with square brackets (*[]*). Both `.iloc` and `.loc` use the indexing operator to receive data from different rows and columns.

Example: Selecting a single column from our project_data.csv file

```
import pandas as pd

dframe = pd.read_csv("project_data.csv",
index_col ="Project Name")

# indexing operator to retrieve columns

firstindex = dframe["Duration"]

print(firstindex)
```

Output

Project Name	Duration
Project A	57
Project B	92
Project C	264
Project D	43
Project E	110
Project F	190
Project G	61
Project H	16
Project I	100
Project J	45
Project K	61
Project L	30
Project M	30
Project N	71

project Name: length:15, dtype:float

If you need to select **multiple columns**, you can modify the operator with the specific columns you need to retrieve:

```python
import pandas as pd

dframe = pd.read_csv("project_data.csv",
index_col ="Project Name")

columns = dframe[["Project Name", "Start Date",
"End Date", "Project Budget"]]

print (columns)
```

Project Name	Start Date	End Date	Project Budget
Project A	5/5/2016	7/1/2016	$1,000,000.00
Project B	5/10/2016	8/10/2016	$900,000.00
Project C	6/10/2016	3/1/2017	$860,000.00
Project D	6/22/2016	8/4/2016	$1,000,000.00
Project E	7/14/2016	11/1/2016	$294,000.00
Project F	7/14/2016	1/20/2017	$123,400.00
Project G	8/1/2016	10/1/2016	$250,500.00
Project H	8/14/2016	8/30/2016	$127,200.00
Project I	9/1/2016	12/10/2016	$80,000.00
Project J	10/1/2016	11/15/2016	$77,000.00
Project K	10/1/2016	12/1/2016	$65,000.00
Project L	11/1/2016	12/1/2016	$550,000.00
Project M	11/10/2016	12/10/2016	$45,000.00
Project N	12/1/2016	2/10/2017	$32,500.00

15 rows * 3 columns

Indexing using *DataFrame.loc[]*

We use the *.loc[]* function to obtain a row or column label. This function operates differently from the indexing operator, as it selects various subsets of rows and columns from the dataset. If you want to select a **single row**, put a single row label inside the square brackets:

```
import pandas as pd

dframe = pd.read_csv("project_data.csv",
index_col ="Project Name")

# retrieve row by loc method

Row2 = dframe.loc["Project B"]

Row3 = dframe.loc["Project C"]

print(row2, "\n\n\n", row3)
```

Project Name		Project B
Start Date		5/10/2016
End Date		8/10/2016
Duration	92	
# of Team Members		2
Project Budget		$900,000.00
Actual Budget		$920,000.00

Project Name: Project B, dtype: object

Project Name		Project C
Start Date		6/10/2016
End Date		3/1/2017
Duration	264	
# of Team Members		4
Project Budget		$860,000.00
Actual Budget		$880,000.00

Project Name: Project C, dtype: object

If you need to **import multiple rows**, put the row labels in a list and call the *.loc()* function:

```
import pandas as pd

dframe = pd.read_csv("project_data.csv",
index_col ="Project Name")

#multiple rows
```

```
rows = dframe.loc[["Project A", "Project B"]]

print (rows)
```

Project Name	Start Date	End Date	Duration	Team Members	Project Budget	Actual Budget
Project A	5/5/2016	7/1/2016	57	10	$1,000,000.00	$880,000.00
Project B	5/10/2016	8/10/2016	92	2	$900,000.00	$920,000.00

If you want to select **more rows and columns**, modify the $.loc[]$ function to fetch data from both rows and columns:

```
import pandas as pd

# Retrieve dataframe csv file

dframe = pd.read_csv("project_data.csv",
index_col ="Project Name")

# retrieving two rows and three columns by loc
method

rowcol = dframe.loc[["Project A", "Project B"],

                ["Start date", "End date",
"Duration"]]

  print(rowcol)
```

Project Name	Start Date	End Date	Duration
Project A	5/5/2016	7/1/2016	57
Project B	5/10/2016	8/10/2016	92

Indexing DataFrame using .iloc[]

.iloc[] allows you to retrieve both rows and columns by position. If you need to select data from a certain row or column, specify the row and column position. This ensures that you extract data from only that selected row or column. The .iloc[] function uses an integer number to specify the location of the row or column you select.

```
import pandas as pd


dframe = pd.read_csv("project_data.csv",
index_col ="Project Name")

# .iloc method to retrieve 3 rows

rows = dframe.iloc[3]

print(rows)
```

Extracting data from multiple rows and columns

```
import pandas as pd

 # Retrieve csv dataframe

dframe = pd.read_csv("project_data.csv",
index_col ="Project Name")

 # .iloc method to retrieve multiple rows

rows = dframe.iloc [[1, 5],[2, 5]]



print (rows)
```

Project Name	# of Team Members
Project A	10
Project B	2

Using *head()* and *tail()* to Peek at Data

We use the `head ()` and `tail()` functions to extract data from the **first rows** and the **last rows** of a DataFrame. By default, the `head()` function returns the **first five rows** of a DataFrame.

Syntax

```
Dataframe.head(n=5)
```

Example

From our **project_data.csv** file, we will return the **top five rows** of the DataFrame and store in a new variable:

```
import pandas as pd

dframe = pd.read_csv("project_data.csv")

top_rows = dframe.head()

print (top_rows)
```

Since the default parameter for $head()$ is 5, you don't have to pass any parameters to the function.

Project Name	Start Date	End Date	Duration	Team Members	Project Budget	Actual Budget
Project A	5/5/2016	7/1/2016	57	10	$1,000,000.00	$880,000.00
Project B	5/10/2016	8/10/2016	92	2	$900,000.00	$920,000.00
Project C	6/10/2016	3/1/2017	264	4	$860,000.00	$880,000.00
Project D	6/22/2016	8/4/2016	43	5	$1,000,000.00	$998,050.00
Project E	7/14/2016	11/1/2016	110	7	$294,000.00	$280,000.00

Calling $head()$ with n^{th} number of rows.

import pandas as pd

df = pd.read_csv("project_data.csv")

n = 9

 # enter data series variable

dataseries = df["Project Name"]

 # return the top 9 rows

top = dataseries.head(n = n)

Print (top)

Project Name
Project A
Project B
Project C
Project D
Project E
Project F
Project G
Project H
Project I

The *tail ()* function works similar to the *head ()* function, as it returns the **last nth number of rows** in a dataframe.

Syntax

```
Dataframe.tail(n)
```

Use the same example and replace `head()` function with the `tail()` function to display the last 9 rows in the dataframe.

Where *n* is the number of rows you want returned.

DataFrame `describe ()` function

The Pandas `describe()` function is a great data analysis tool that enables you find summarized data. The function can perform basic statistical calculations like calculating **mean**, **STD**, and **percentile**, among others.

A **percentile** is the nth value in a group of population that is divided into two: the lower value and the upper value. The lower value contain values of n percent of where the data falls whereas the upper part contains the rest of the data. It indicates a value below which a given percentage of observation in a dataset falls. For example, a percentile of 30 is the value in which 30% of the observations fall below that value.

Percentiles are represented by values between 0-100 with 50th percentile being the median.

Syntax

```
DataFrame.describe(percentiles=None, include=None, exclude=None)
```

Where,

percentiles=: shows list-like datatype of numbers between 0-1 to return the percentile.

include=: displays the list of datatypes to include when describing the DataFrame. Default datatype is **none**.

exclude=: displays the list of datatypes to exclude when describing the DataFrame. Default datatype is **none**.

Return type: the statistical summary of the dataframe.

Example

Employee_details.csv is a file with a list of 364 employees with information about their relevant details. It is not included in this book since it would take up too much space; and it does include personal information that we do not want to disclose.

We will create a statistical summary of a DataFrame with both objects and a numeric data type. Using the **Employee_details.csv** file, we will calculate the respective percentile for the object series **[.20, .40, .60, .80]**.

import pandas as pd

```
import pandas as pd
# call function to regex module
import re
data = pd.read_csv ("Employee_details.csv")

# remove null values and set inplace value to true to catch
errors
data.dropna (inplace = True)
perc = [.20, .40, .60, .80]
  # dtypes to retrieve
include = ['int', 'float', 'object']

desc = data.describe (percentiles = perc, include = include)
desc
```

	Name	Team	Number	Position	Age	Height	Weight	College	Salary
count	364	364	364.000000	364	364	364	364.000000	364	3.640000e+02
unique	364	30	NaN	5	22	17	NaN	115	NaN
top	Cleanthony Early	New Orleans Pelicans	NaN	SG	24.0	6-9	NaN	Kentucky	NaN
freq	1	16	NaN	87	41	49	NaN	22	NaN
mean	NaN	NaN	16.829670	NaN	NaN	NaN	219.785714	NaN	4.620311e+06
std	NaN	NaN	14.994162	NaN	NaN	NaN	24.793099	NaN	5.119716e+06
min	NaN	NaN	0.000000	NaN	NaN	NaN	161.000000	NaN	5.572200e+04
20%	NaN	NaN	4.000000	NaN	NaN	NaN	195.000000	NaN	9.472760e+05
40%	NaN	NaN	9.000000	NaN	NaN	NaN	212.000000	NaN	1.638754e+06
50%	NaN	NaN	12.000000	NaN	NaN	NaN	220.000000	NaN	2.515440e+06
60%	NaN	NaN	17.000000	NaN	NaN	NaN	228.000000	NaN	3.429934e+06
80%	NaN	NaN	30.000000	NaN	NaN	NaN	242.400000	NaN	7.838202e+06
max	NaN	NaN	99.000000	NaN	NaN	NaN	279.000000	NaN	2.287500e+07

97

From the output, you can see that for all the percentile and columns with strings, a **NaN** is returned.

Example 2: Using `describe()` in a series of strings

```
import pandas as pd
# importing regex module
import re
data = pd.read_csv("Employee_details.csv")
data.dropna(inplace = True)
#use describe() to retrieve employee name
desc = data["Name"].describe()
desc
```

When you run the program, it will return unique values and count all the values and the frequency of occurrence, as shown below.

```
Using describe() to count values

count                          364
unique                         364
top            Cleanthony Early
freq                             1
Name: Name, dtype: object
```

DataFrame *Resample()* function

Resample() function is essential for data manipulation. When you have a dataset recorded in a set time interval, you can change the time interval using **the resampling technique**. You may set the time interval to daily, weekly, monthly, or semiannually. You can also resample hourly data into minute-by-minute data.

The function is ideal for frequency conversion and for sampling time series data. In time series data, you would index the data points in a particular time order; thus, *resample()* is a convenient method for resampling time series data.

All objects in the *resample()* method should have a **datetime index** (either a DatetimeIndex, PeriodIndex, or a TimedeltaIndex) or they should pass values in the form of datetime.

Syntax

*DataFrame.**resample(arguments).<aggregate function>***

Resample arguments:

```
DataFrame.resample(rule, how=None, axis=0, fill_method=None,
closed=None, label=None, convention='start', kind=None,
loffset=None, limit=None, base=0, on=None, level=None)
```

Where,

rule: = the offset string or object that represents the **target conversion**. The parameter you use determines the interval by which the program will resample data:

```
data.resample('3min').sum()
```

axis: contains data of *int* type, which is optional. Default value= 0

closed: {'right', 'left'}

label: {'right', 'left'}

convention: for PeriodIndex only; it controls whether to use the start or end of the rule.

loffset: a parameter used to adjust the resampled time labels.

base: a parameter for frequencies that evenly subdivide one day; the "origin" of the aggregated intervals. For example, for a "5 min" frequency, the base can range from 0 to 4. Default value= 0.

on: parameter set for column DataFrame instead of index for resampling. Column must be **datetime-like**.

level: For a MultiIndex level (name or number) to use for resampling. Level must be **datetime-like**.

Types of Resampling

Resampling is categorized into two: **downsampling** and **upsampling**.

Downsampling: where the function resamples data from a wider time frame, such as months or years. When downsampling, you can use several groups by aggregating functions like `min()`, `max()`, `sum()`, `means()`, etc.

When you downsample, you reduce the number of rows in the dataframe.

Upsampling: allows you to resample your data for a short period.

Resampling enables you to create a unique distribution of data. For example, if you use time series data with six-minute timestamps, you can *downsample* the time series into three minutes.

Example

We will use the **shares_value.csv** file and *resample()* the data on a **monthly basis**. You can download a sample CSV file from this <u>link.</u>

```python
import pandas as pd

# use parse_date function to convert timestamp
column into datetime format.

 parse_dates =["Timestamp"]

# Resampling works well with time-series data
therefore, convert timestamp column to an index
form

 index_col ="Timestamp"

data =pd. read_csv("shares_value.csv",
parse_dates =["Timestamp"], index_col
="Timestamp")

 # Print 12 rows of the shares_value dataframe

data[:12]
```

Timestamp	closing value	volume	open	high	low
12/20/2018	192.23	46541444	191.72	197.18	191.4501
12/13/2018	192.23	46725710	191.63	197.18	191.4501
12/12/2018	194.17	50991030	199	199.85	193.79
12/9/2018	204.47	34317760	205.55	206.01	202.25
11/8/2018	208.49	25289270	209.98	210.12	206.75
11/7/2018	209.95	33291640	205.97	210.06	204.13
11/6/2018	203.77	31774720	201.92	204.72	201.69
11/5/2018	201.59	66072170	204.3	204.39	198.17
11/2/2018	207.48	91046560	209.55	213.65	205.43
11/1/2018	222.22	52954070	219.05	222.36	216.81
10/31/2018	218.86	38016810	216.88	220.45	216.62
10/30/2018	213.3	36487930	211.15	215.18	209.27

With the above data, you can calculate the **mean closing price** for each month.

```
import pandas as pd


parse_dates =["timestamp"]
```

```
 index_col ="timestamp"

data =pd.read_csv("shares_value.csv",
parse_dates =["timestamp"], index_col
="timestamp")

#create variable M to represent months and
calculate mean


monthly_resampled_data=data.close.resample('M').
mean()


# Print closing mean for 12 months

data[monthly_resampled_data]
```

```
Resampled mean() for 12 months
date
2017-11-30    172.090769
2017-12-31    171.891500
2018-01-31    174.005238
2018-02-28    167.638947
2018-03-31    174.496190
2018-04-30    169.834286
2018-05-31    185.536818
2018-06-30    188.621429
2018-07-31    190.311429
2018-08-31    213.346087
2018-09-30    222.073684
2018-10-31    220.845652
2018-11-30    204.930000
Freq: M, Name: close, dtype: float64
```

Graph for frequency conversion

import matplotlib.pyplot as plt

plotly.data[monthly_resampled_data]

plt.show()

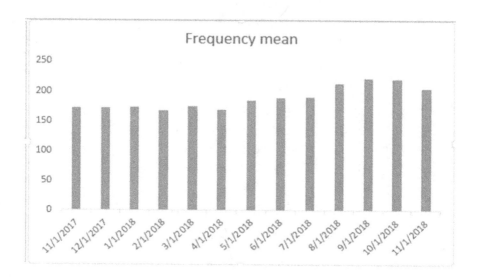

When resampling data on a **weekly basis**, you can apply the mean opening price to the dataframe.

The code will look like this:

```
Weekly_resampled_data=df.open.resample('n').mean()
```

Chapter Summary

In this chapter, you learned how to work with time series data, with examples. You also learned how to use the `Pandas_datareader()` function to import data into the workspace and carry out statistical functions for analyzing the data. Some of these statistical functions included:

- Using *head()* and *tail()* functions to peek data

- Using *describe()* to generate summary data

- Using *resample()* function for frequency conversion and sampling of time series data

- How to use *.loc()* and *.iloc()* to select data from different columns and rows.

In the next chapter, we will look at how to visualize data using time series, and how to plot time series charts and scatter matrix.

How To Visualize Time Series Data

What is Data Visualization?

Data visualization is a presentation of data in the form of charts or graphs. The plotted graphs make it easy to understand, allowing you to detect trends, patterns, and outliers easily in any group of data.

Data visualization is essential for data analysis and observation of time series data. Over recent years, more companies have been investing in big data and machine learning to analyze large volumes of collected data. Data analysts require tools that will enable them to comprehend and present complex organization data for stakeholders and business owners.

One way to make data more presentable is to convert the complex data into a visual format as compared to having multiple lines of text and numbers.

Data visualization not only saves you the time of going through multiple lines to understand data, but it also makes it easy for you to identify trends in the optimization process.

Python provides you with the tools you need to present visual data. It offers multiple graphical libraries that provide wide features for plotting graphs and charts. You can easily create highly interactive graphics or customize charts based on your needs.

Some of these Python libraries for data visualization include:

Matplotlib: This is a low-level library that gives the freedom to customize your charts based on your needs. To install Matplotlib, you can use both *pip* or *conda*.

```
pip install matplotlib
```

or

```
conda install matplotlib
```

You can use Matplotlib when you want to create basic charts like bar charts, line charts, histograms, and pie charts, among other basic charts.

To use Matplotlib, import the library into your workspace.

Syntax

```
Import Matplotlib.pyplot as plt
```

Pandas Visualization: This is a great, easy-to-use, open-source tool with a high-performance library that can provide lots of data structures. These structures include Dataframes and data visualization tools to analyze and present data. It is built on the Matplotlib library and makes it easy to create plots from stored Pandas DataFrame or series.

Because of its high level of API compared to Matplotlib, the Pandas visualization tool requires less coding.

You can install Pandas through:

```
pip install Pandas
```

or

```
conda install Pandas
```

Seaborn: This is a great visualization tool for high-level interfaces and has plenty of default styles. Using the seaborn library, you can create a graph that can read data from multiple lines in Matplotlib. It has an interactive interface that gives an opportunity to work with Pandas DataFrames.

To use this library, import it to your IDE interface.

```
import seaborn as sns
```

Plotly: Plotly is an advanced data visualization tool that includes the Python API libraries. The tool integrates with Python to make interactive web-based apps. Install the library to your device by running the *pip* or *conda* install.

```
pip install plotly
```

or

```
conda install plotly
```

Once you install plotly, you have to authenticate the library and replace the username and API key.

```
import plotly.plotly as py

# Authenticate account with your username and password
plotly.tools.set_credentials_file(username='enter username',

                                   api_key='enter password')
```

If you plan to use plotly online, you need to create an account with a username. A free plotly account allows you to create 25 public charts and one private chart; therefore, if you want to create more charts, you have to delete the old charts to create more space for new ones.

You can also set up an offline plotly library with the following code:

```
import plotly.plotly as py
import plotly.graph_objs as go

# Offline plotly mode
from plotly.offline import init_notebook_mode, iplot
init_notebook_mode(connected=True)
```

You can easily create basic plotting techniques using Matplotlib, Pandas visualization, and the seaborn library. You can also learn how to use the various features to change your chart's color, grid, and scale. In the following sections, we will focus on using the Matplotlib package.

Plotting Simple Data with Matplotlib

Plotting a graph is very easy. We will look at an example of plotting your first line graph.

```python
import matplotlib.pyplot as pt



# x axis values

x = [1,4,5]

# y axis values

y = [2,3,1]



# plotting the points

pt.plot(x, y)
```

```python
# naming the x axis

pt.xlabel('x - axis')

# naming the y axis

pt.ylabel('y - axis')


# Enter graph title

pt.title('Welcome to the first graph!')


# function to show the plot

pt.show()
```

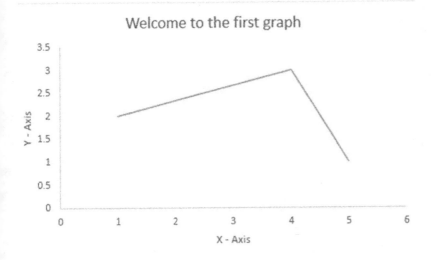

Welcome to the first graph

We use the `.plot ()` function to plot the graph on the canvas.

You can use both the `.xlabel ()` and `.ylabel ()` functions to name your *x*- and *y*-axes.

The `.title ()` function allows you to title the graph.

The `.show ()` function displays the graph.

Customizing Your Plots

You can customize the graph to your needs, including the plotting point color, style, and sometimes even the scale.

```
import matplotlib.pyplot as plt
#create x and y variables
x = [1,2,3,4,5,6]
y = [2,4,1,5,2,6]
  # plotting the points
plt.plot(x, y, color='green', linestyle='dashed', linewidth = 3,
        marker='o', markerfacecolor='blue', markersize=12)
 plt.ylim(1,7)
plt.xlim(1,7)
plt.xlabel('x - axis')
plt.ylabel('y - axis')
plt.title('Graph customization!')
  # show the plot
plt.show()
```

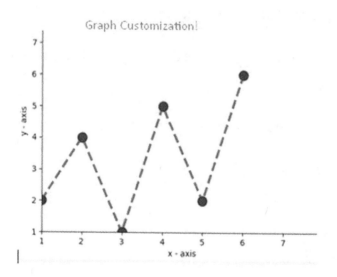

Adding a Background Grid to Your Plot

```python
import plotly.plotly as py
import plotly.graph_objs as go
import plotly.tools as tls
import matplotlib.pyplot as plt
import numpy as np

grid = plt.grid()
x = np.linspace(0, 14, 100)
gflip= -1
for i in range(1, 6):
    plt.plot(x, np.sin(x + i * .5) * (6 - i) * gflip)

plotly_grid = tls.mpl_to_plotly(grid)
plotly_grid['layout']['x-axis']['showgrid'] = True
plotly_grid['layout']['y-axis']['showgrid'] = True

plotly_url = py.plot(plotly_grid, filename='background-grid')

plt.show()
```

Changing Graph Grid

Changing Axis Label Font Size

```python
import matplotlib.pyplot as plt
import numpy as np
import plotly.plotly as py
import plotly.tools as tls


mpl_grid = plt.grid()
label = mpl_grid.add_subplot(111)

t = np.arange(-5.0, 5.0, 0.01)
s = np.sin(4 * np.pi * np.absolute(t)) * np.exp(-5 *
np.absolute(t))
line, = label.plot(t, s, lw=2)

label.set_title("My plot title")
label.set_xlabel("X-Axis fontsize=18", fontsize=18)
label.set_ylabel("Y-Axis fontsize=16", fontsize=16)

plotly_grid = tls.mpl_to_plotly( mpl_grid )

plot_url = py.plot(plotly_grid, filename='mpl-axes-
font-size')
```

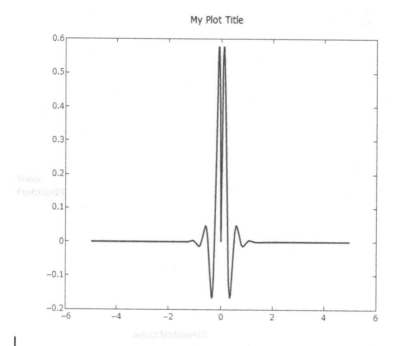

My Plot Title

Visualizing Time Series Data

Time series data is a sequence of discrete-time data taken over successive, equally spaced points in time. A collective observation of sales data for one month represents a time series data. This time series data defines and consistently measures the sales revenues at successive, equally spaced points in time.

Time series data comprises three components:

- **The trend** (activities with long-term direction)

120

- **Seasonal** (systematic or calendar-related movements)

- **Irregular** (unsystematic or short-term fluctuations)

Time series can either be in the form of *stock* or *flow*. A **stock** series measures data attributes at a certain point in time, which investors would commonly view as "stocktales." For example, employee payroll data is a monthly stock measure that determines whether the employee has rights to any payments that month.

A **flow** series is a measure of activities that have taken place over a given time period. An example is the survey of manufacturing activity for a week. Every day, the company would produce a specific amount of products within a week; they can sum up the amounts to give a total value of production for that week.

Example

Time series data is a sequence of discrete-time data taken over successive, equally spaced points in time. A collective observation of sales data for one month represents a time series data. This time series data defines and consistently measures the sales revenues at successive, equally spaced points in time.

Time series data comprises three components:

- **The trend** (activities with long-term direction)

121

- **Seasonal** (systematic or calendar-related movements)
- **Irregular** (unsystematic or short-term fluctuations)

Time series can either be in the form of *stock* or *flow*. A **stock** series measures data attributes at a certain point in time, which investors commonly view as "stocktales." For example, employee payroll data is a monthly stock measure that determines whether the employee has rights to any payments that month.

A **flow** series is a measure of activities that have taken place over a given time period. An example is the survey of manufacturing activity for a week. Every day, the company would produce a specific amount of products within a week; they can sum up the amounts to give a total value of production for that week.

Example

In this example, I will use the shares value data, you can download a sample file to use from this link. This data shows the sales volume and closing stock prices of shares for a period of two years.

Visualizing this data will enable you to understand the shares trends and when shares value is high, low or even open prices. You need to analyze the trends before making any investment decision.

Timestamp	closing value	volume	open	high	low
12/20/2018	192.23	46541444	191.72	197.18	191.4501
12/13/2018	192.23	46725710	191.63	197.18	191.4501
12/12/2018	194.17	50991030	199	199.85	193.79
12/9/2018	204.47	34317760	205.55	206.01	202.25
11/8/2018	208.49	25289270	209.98	210.12	206.75
11/7/2018	209.95	33291640	205.97	210.06	204.13
11/6/2018	203.77	31774720	201.92	204.72	201.69
11/5/2018	201.59	66072170	204.3	204.39	198.17
11/2/2018	207.48	91046560	209.55	213.65	205.43
11/1/2018	222.22	52954070	219.05	222.36	216.81
10/31/2018	218.86	38016810	216.88	220.45	216.62
10/30/2018	213.3	36487930	211.15	215.18	209.27
10/29/2018	212.24	45713690	219.19	219.69	206.09
10/26/2018	216.3	47191700	215.9	220.19	212.67
10/25/2018	219.8	29027340	217.71	221.38	216.75
9/24/2018	215.09	39992120	222.6	224.23	214.54

From the shares_value data, we can plot a single column data

```
import pandas as pd



# Read in data with two headers

data = pd.read_csv('shares_value.csv',
header=[0,1], index_col=0)



# Retrieve column data
```

```
sales = df.loc[:, ('timestamp', 'sales volume')]


# Plot the data in a chart

sales.plot()
```

From the graph, notice that sales are always low at the beginning of the year but increase during the year, with high sales recorded at the end of the year. This represents an upward trend within the year, although there are months where sales are low.

Using a Scatter Matrix Data Visualization Tool

A **scatter plot** or **scatter matrix** is a collection of scatter plots organized to form a grid. The grid shows a relationship between a pair of variables where you would plot each variable against the other. For example, if you have k variables, your scatter matrix will have k rows and k columns.

A scatter plot is a powerful data visualization tool that conveys detailed information. The chart represents a two-dimension data visualization with dots generated from two variables. The program plots the data against the x- and y-axes.

For example, in a graph representing the **weight of children**, each dot will represent a **single child with his or her height**. The graph will measure the children's *height* along the x-axis, while representing the children's *weight* along the y-axis.

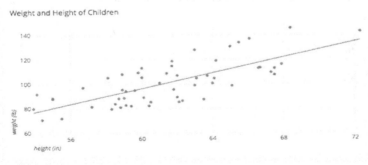

125

When to Use Scatter Charts

Scatter plots are useful when you want to show the relationship between two variables. They determine a **linear correlation** between different variables. In the above example, the chart not only shows the weights and heights of children but also visualizes the relationship between the weight and the height. As the height increases, so does the weight; therefore, we can say that height correlates to the weight of each child.

If there is no correlation between the variables, the chart will display the variables as scattered on the coordinate plane. In case of a large correlation, the dots will concentrate near the straight line.

If the y-axis variable *increases* as the x-axis variable *increases*, then the correlation is **positive**; if the y-axis variable *decreases* when the x-axis variable *increases*, it is a **negative** correlation.

If you can't tell whether y-axis is increasing or x-axis is increasing (and vice versa), then you have **zero correlation**.

A positive correlation has a maximum of +1 and 100% when the dots lie exactly along the straight line. A negative correlation has a maximum of -1 and 100%.

Types of Scatter Charts

There are three types of scatter diagrams based on the correlation.

- No Correlation

- Moderate Correlation

- Strong Correlation

Scatter Diagram with No Correlation

This scatter chart has a **zero correlation** because the dots are spread randomly and you can't draw any line between them. There is no relationship between the two variables.

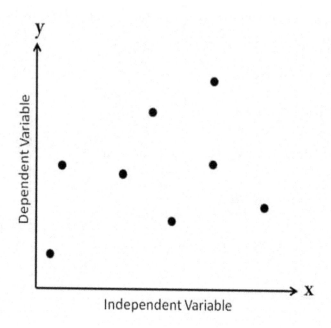

Scatter Diagram with Moderate Correlation

We can also call this chart a scatter diagram with **low degree of correlation**.

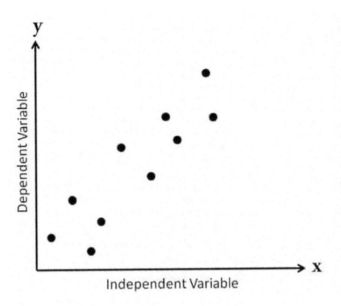

The data points in this chart somewhat close together, with some correlation between the two variables.

Scatter Diagram with Strong Correlation

This scatter plot has a high degree of correlation because the data points grouped close together. You can easily draw the line following the data points.

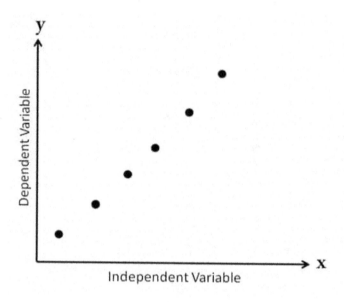

This scatter chart shows the variables as closely related to each other.

When drawing scatter plots, you can classify them based on the trend and the slope of the data points.

Advantages of Drawing Scatter Plots

- Plotting the chart is relatively easy.

- Observation and reading the plotted data is straightforward.

- It is the best tool to show you the relationship between two variables.

- The chart can easily show you non-linear data patterns.

- It shows the range of data flow; you can determine maximum and minimum data values.

How to Plot the Scatter Matrices

Here, we will create a **simple scatter plot** using matplotlib *scatter ()* method.

```
import matplotlib.pyplot as pt


# x-axis values

x = [2,3,4,5,4,5,7,8,10,12]

# y-axis values

y = [1,2,3,6,5,6,7,8,9,10]


# plotting points as a scatter plot
```

```python
pt.scatter(x, y, label= "series", color= "blue",
            marker= "bo", s=32)


pt.xlabel(' x - axis')


pt.ylabel(' y - axis')
# plot title
pt.title(' Scatter Diagram!')
pt.legend()


# function to show the plot
pt.show()
```

Scatter Matrix Using Pandas Library

Creating a scatter matrix will help to visualize data trends. Pandas library uses a Matplotlib package to display the scatter matrix.

When creating a scatter plot in Pandas library, call the `<dataset>.plot.scatter()` function and pass the arguments to the function (for the x-axis column and that for the y-axis column).

Example: Plotting Scatter Matrix Using Shares_value.csv Data

```
import matplotlib.pyplot as pt

import pandas as pd

from sklearn import datasets
```

```
%matplotlib inline


pt.style.use('style')

#load data to plot

data = pd.read_csv (shares_value.csv)

data = pd.DataFrame('Timestamp', 'open', 'low',
'high')

pd.scatter_matrix(data, alpha=0.2)

pt.show()
```

In the above graph, you learnt that there is a close correlation between low, open and high values daily. So you can use one of those values to determine trends in stock prices.

Chapter Summary

Data visualization is an essential tool for data presentation. It enables you to display data in a simple and easy-to-understand manner. Various data visualization tools like charts, graphs, and other representations of data in a summarized form. This helps to grasp information from mass volumes of data within a few minutes.

In this chapter, you learned how to visualize data using various tools, including visualizing time series data. You also learned how to plot data using Matplotlib package and data visualization using Pandas library.

We went over how to plot different charts including scatter matrices and how to change chart color, grid, and scale. In drawing scatter diagrams, we can now highlight the essentials for using scatter plots for data presentation and how you can read the data points to determine the relationship between two variables.

In the next chapter, you will learn about common financial analysis tools you can use to analyze your data for future decision-making.

A Short message from the Author:

Hey, are you enjoying the book? I'd love to hear your thoughts!

Many readers do not know how hard reviews are to come by, and how much they help an author.

I would be incredibly thankful if you could take just 60 seconds to write a brief review on Amazon on the product page, even if it's just a few sentences!

Thank you for taking the time to share your thoughts!

Your review will genuinely make a difference for me and help gain exposure for my work.

Common Financial Analysis

What is Financial Analysis?

Financial analysis is a process used to examine the financial data of an organization to make informed decisions. It is an assessment tool to determine how profitable a venture is in terms of its income statement, cash flow, and its financial position. Financial analysis on any project undertaken by the organization determines how viable it is and recommends the management team on areas they need to improve.

Many industries use Python in finance for qualitative and quantitative analysis. They also use Python in stock market analysis, prediction, and machine learning.

There are powerful libraries in Python like Pandas, Numpy, and Anaconda that organizations use for financial data analysis and other calculations. These libraries can easily integrate with other platforms and languages to perform specific functions.

The Pandas library plays a larger role in working with data. It helps to analyze large volumes of data and visualization, thus making it easy to perform statistical calculations.

Libraries like Scikit-Learn and PyBrain are useful in machine learning and prediction analysis.

Trading Financial Analysis

Traders require a lot of analysis in the stock market, and they commonly use Python to handle analysis more often than other programs. The program makes it easy for developers to define winning trading strategies and also gives some tips on stock trading patterns in the market.

Python development framework makes it easy to examine stock information in the stock exchange market and make an informed decision. Investors can also get some recommendations on open stock opportunities they can invest in based on the analyzed data.

Information like stock prices and current market prices is available to investors to make the right decisions on what to invest in and when. Stock prices change daily, which can affect the value of an investment, thus the need for carrying stock analysis and visualization.

Financial analysis of the stock trade enables you to measure your daily gains or losses especially if you have invested in long term stocks. These stocks act as a way of generating passive income since you will receive a share of profits from the company proceeds.

The goal of any trading financial analysis is to maximize returns for the investors despite stock constraints.

Some of these constraints include;

- Individual ethical concerns,

- Current income needs of an individual and,

- Risk involved

Stock prices change on a daily basis, therefore, there is a need to monitor the daily changes and calculate the daily returns based on your investment. Monitoring the daily changes will enable you to know the magnitude of change and predict the future changes in the stock prices.

Daily stock returns also helps in estimating your future total returns on all stock investment. It is important to determine daily return for each stock investment. Your future returns will be based on dividends paid and changes in the share price.

Share price changes can be in the form of change in per-share-earnings and change in the price-to-earning on multiple stocks.

Daily Returns

Analysing daily stock returns helps to monitor its rate or magnitude of change. It helps measure the dollar change in the stock price and it is expressed as a percentage of the previous day stock closing price.

If the return is positive, the stock price grows in value, whereas a negative return shows a loss in value. Stocks with higher daily returns are perilous compared to stocks with lower positive and negative daily returns.

To calculate your daily stock return, you need to compare the closing price of the previous day with the current stock price and express the difference between the prices into a percentage value.

You can calculate the daily return value manually or use an online calculator for quicker results. You can also develop a Python code that will automatically generate your daily return based on a few parameters.

How to Determine Daily Return

To know your daily return, you need to first know how much the stock is worth. To do this, visit financial websites that give daily stock price information. Search stock prices of the specific company you're interested in from the website search button.

Once you get the results, you can get the daily return by dividing the **current price** by the **previous day's closing stock price**, then multiply by 100.

```
Daily Return = [Current stock price/previous
closing stock]*100
```

Assume the current closing stock is $2.25 and the previous closing stock was $25.5 then,

Return for the day =[$2.25/$25.5]*100 to get **8.8%**

To calculate the gain or loss, subtract the **current stock price** from the **previous closing price**. Multiply the results by 100 to convert it to a percentage, then multiply by the **number of shares you owe**.

```
Actual gain/loss = [(Previous closing stock-Current stock
price)*100]*shares
```

You can also convert your daily return to annual return. The return per day is expressed as a decimal number.

```
Daily investment return = [Amount of return / value of the
investment] *100
```

$$Annual\ return\ =\ [(Daily\ return+1)^{365} -1]*100$$

To convert our daily return to annual return, we raise the figure to the power of 365 (*days*), then subtract by 1. We can convert the value we get into percentage by multiplying with 100.

For example, if you have an investment that gives you 0.06% daily return, convert this figure into decimal format and convert the figure to annual returns using the formula:

$$Annual\ return=\ [(0.0006+1)^{365}-1]*100=24.5\%$$

Calculating Daily Return via Python Pandas.

To use Pandas library, you need to import all the modules to your Python library.

```
import Pandas as pd
import numpy as np
import matplotlib.pyplot as plt
import Pandas_datareader as web
```

After importing the modules, you can download historical prices of data from the Internet and use the downloaded prices to calculate the returns.

Example

In this example, we will download **stock data** that shows the stock performance for a certain period, then use those historical prices to perform the returns.

```
import pandas as pd

import numpy as np

import matplotlib.pyplot as pt

import pandas_datareader as web

from pandas_datareader import data as pdr

sp= web.get_data_yahoo("NFLX"),

start_sp= datetime.datetime ("2009/01/01"),

end_sp =datetime.datetime('2018/03/01'),



print(sp.head())
```

when you run the program, this is the output you will be able to obtain.

Stock Performance for the top 5 stocks

	High	Low	...	Volume	Adj Close
Date			...		
2009-01-02	4.357143	4.200000	...	6605200.0	4.267143
2009-01-05	4.562857	4.302857	...	13044500.0	4.562857
2009-01-06	4.750000	4.590000	...	12065900.0	4.705714
2009-01-07	4.734286	4.571429	...	10133900.0	4.672857
2009-01-08	4.797143	4.485714	...	8175300.0	4.735714

To obtain the **closing stock**, run this code:

```
sp['Adj Close'].plot()

#name the axis

pt.xlabel("Dates")

pt.ylabel("Adjusted closing price")

pt.title("Share prices data")

pt.show()
```

Calculating Daily Return

To calculate the daily return, we use pct_change() function

```
stock_daily_returns = sp['Adj
Close'].pct_change()



print (stock_daily_returns.head())
```

Output

```
       Daily Return Based on Adj Closing stock
Date             Adj Closing Stock
2009-01-02            NaN
2009-01-05       0.069300
2009-01-06       0.031309
2009-01-07      -0.006982
2009-01-08       0.013452
Name: Adj Close, dtype: float64
```

Insert daily return chart for stock price data:

```
#Set the graph size and axis

fig = pt.figure()

pt = fig.add_axisrange([0.1,0.1,0.8,0.8])

pt.plot(stock_daily_returns)

pt.set_xlabel("Date")

pt.set_ylabel("Percent")

pt.set_title("Daily returns data")

pt.show()
```

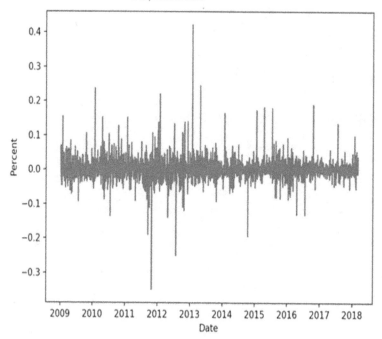

Cumulative Daily Rate of Return

You would use the cumulative return to show the total amount of return an investor can receive from his or her investment, irrespective of the time. It is the aggregate amount lost or gained in an investment independent of the time involved. You would always express the amount as a percentage.

Coming up with the daily returns will make it easy to plot the volatility of the investment. If you have been investing in stocks for a while, you can determine the growth of your investment by calculating cumulative returns.

To obtain the cumulative return in Python, use the *cumprod ()* function:

```
Sp_cum_returns = (Sp_daily_returns +
1).cumprod()
```

Using data from the previous example about daily returns, we can plot a **graph of daily cumulative returns**:

```
#prepare graph size

fig = pt.figure()

pt = fig.add_axisrange([0, 70])

sp_cum_returns.plot()

pt.set_xlabel("Date")

pt.set_ylabel(" Investment")

pt.set_title("Cumulative daily returns data")

pt.show()
```

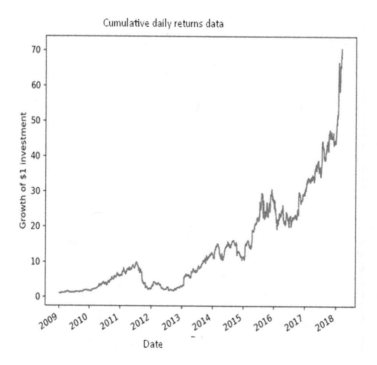

Cumulative daily returns data

Moving Window

In finance, we would call evaluating statistical trading metrics over a sliding window time frame as a moving window. A moving window means isolating subsets of a time-dependent measurement through taking the last *n* time segment and using each time segment as an input in deciding the best trading strategy. It analyzes price history by averaging the daily stock prices over time. Many investors and organizations use moving windows to determine moving averages of stock prices.

A moving window acts as a technical analysis tool that smoothes out share prices by filtering out "*noise*" from the short-term price fluctuations. The technique uses past share prices and can easily identify the changing trends in prices and also determine support and resistance levels.

Moving averages show the lag in the current market prices as compared to past prices. The lag is directly proportional to the period; the longer the time, the greater the lag. The length of the moving window depends on the trading objectives. Here, investors use short moving averages for short-term trading.

If the moving averages of stocks are rising, then the security is in the uptrend whereas if the moving averages are reducing, they're indicated by a downtrend.

Moving averages are fully customizable and an investor can choose which time-frame they want to use when creating the averages. When finding the best time frame to use, you need to experiment with different time periods and choose the time that fits your strategy.

A moving average trend creates a constantly updated price average through smoothing out the price trends.

How to Calculate Moving Window

The moving average forms the basis of **momentum-based trading strategy**. You can get the rolling mean over a window period. When using the `rolling()` method with a window of twenty days, you can obtain the moving average.

Example 1:

Using `rolling ()`

```
import pandas as pd

# change the adjusted closing prices to
adj_prices

adj_price = msft_data['Adj_Close']


# calculate the moving average
wdays=20
```

```
MA =
pd.series((data['adj_price'].rolling(window=wday
s).mean())

Data=data.join(MA)

# print the result

print(MA)
```

Example 2

To calculate the moving window for the last 100 windows (*days*), we need to find the **closing stock** and also the **average of each window's moving average**. We can plot the moving average against the windows using the Matplotlib.

Matplotlib will enable you to plot the moving average against the stock price chart.

We will import Apple stock prices from 2011 to 2017 and calculate the moving average:

```
#extract Stocks Price from the internet using the following codes:

import pandas as pd

import datetime
```

```
import pandas_datareader.data as web

from pandas import Series DataFrame


stock_start = datetime.datetime(2011, 1, 5)

stock_end = datetime.datetime(2017, 1, 5)


dframe = web.DataReader("AAPL", 'Apple', stock_start, stock_end)

dframe.tail()
```

From the above data, we can calculate the **moving average**:

```python
#code to calculate moving average with moving window
%matplotlib inline
#import matplotlib to plot the graph
import matplotlib.pyplot as plt
from matplotlib import style

# set matplotlib size
import matplotlib as mpl
mpl.rc('plot', size=(9, 8))
mpl.__version__

# setting matplotlib style using seaborn library
style.use('ggplot')

# set rolling window to 100
close_ad=dframe['Adj Close']
nav=Close_ad.rolling(window=100).mean()

label = plot.add_axes([0,20,40,60, 80, 100)
label.set_xlabel("Date")
label.set_ylabel("Window")
label.set_title(" Moving Average")
close_ad.plot(label='AAPL')
m-avg.plot(label='m-avg')
plt.legend()
```

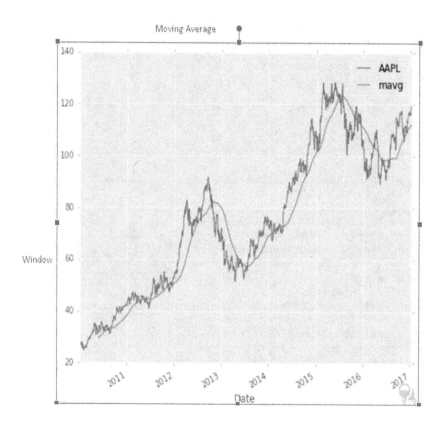

From this graph, you can see how the prices increase and decrease along the moving window.

Volatility Calculations

Volatility is a variation of stock prices over a specified period; a statistical measure of stock dispersion returns over this period. Some stock prices may change at a steady rate within a day, whereas other stock types may move at a higher or lower rate. These stocks may end up being offered at the same price despite their varying paths throughout the day.

Stock or asset volatility helps you measure the risk you're taking in any single investment. You should compare particular stock volatility with another stock to minimize the risk involved. Stocks with high volatility have a higher risk compared to investing in stocks with low volatility.

Calculating stock volatility is very easy, and it mostly depends on the size of the window (*number of days*). We would determine each stock volatility by the rolling-window standard deviation from the percentage change in the stock. The size of the stock window influences the overall results of the stock prices.

If the window is wider, there will be less representative of the stock measurement. When the window narrows, it makes it easy to obtain the standard deviation.

Example

If you want to invest in **Apple's stock securities**, you can first download **Apple's historical data** and use the data to determine the risk of your investment.

To calculate the volatility of your asset, you need to import the Numpy library.

```
import numpy as np

import matplotlib.pyplot as plt

# Define the minimum of periods to consider

min_periods = 75

# Calculate the volatility

vol =
daily_pct_change.rolling(min_periods).std() *
np.sqrt(min_periods)

# plot volatility

Vol.plot(figsize=(10,8))

Plt.show()
```

This example determines the volatility of a single investment (*Apple stock volatility*). If you need to compare the risk of investment in different stocks, you need to **diversify** your investment portfolio and carry out risk volatility for various stocks. Diversification also helps lower the volatility of an investment portfolio.

Dividends

Shareholders benefit from the wealth of the company by receiving dividends in the form of shares. Dividend payments encourage shareholders to continue investing in the company.

Companies give each shareholder additional shares; therefore, the shareholders can have healthy returns on their investments. The company doesn't have to part ways with its capital in the form of payout to the shareholders since they give the latter more shares instead of cash. Paying cash dividends reduces the company's value by the amount paid to the shareholders.

They issue dividends based on the number of shares each shareholder holds. For example, if the company issues dividends by 20%, each shareholder will have their stock holdings increase by 20%.

To calculate the dividend, you take the annual dividend and divide by the current stock price.

Dividend= Annual Dividend/Current Stock Price

For example, if Apple stock had share price of $50 and the annual dividend is $1.00 the dividend yield will be $0.02 or 2%.

You can use **dividend Assistant** tool to estimate your future dividend income. With the tool, you can link to your brokerage account or alternatively add all your stock holdings manually and track all your dividends.

dividend Assistant tool is a free dividend income management tool which can be used at any time. You can download it for free from www.dividend.com.

The table below shows the top rated dividend stocks from dividend.com with the stocks rated using the Dividend Advantage Rating System (DARS).

Stock Symbol	Company Name	DARS™ Rating	Ex-Div Date	Pay Date	Div Payout	Qualified Dividend?	Stock Price	Yield
MCR	MFS Charter Income Trust Shs Ben Int		2019-10-15	2019-10-31	0.06	Unknown	$8.36	8.67%
MGF	MFS Government Markets Income Trust Shs. Ben. Int.		2019-10-15	2019-10-31	0.03	Unknown	$4.61	7.58%
CXE	MFS High Income Municipal Trust Sh Ben Int		2019-10-15	2019-10-31	0.02	Unknown	$5.45	4.51%
CMU	MFS High Yield Municipal Trust Sh Ben Int		2019-10-15	2019-10-31	0.02	Unknown	$4.67	4.63%
CIF	MFS Intermediate High Income Fund Sh Ben Int		2019-10-15	2019-10-31	0.02	Unknown	$2.59	9.44%

Ordinary Least Squares Regression

Ordinary least squares (OLS) regression is a statistical analysis technique used to determine the relationship between independent and

161

dependent variables. It measures the number of squares between the observed and predicted values of the dependent variable.

In the financial world, they often refer to this technique as the **linear regression model**. We use the OLS model to estimate, interpret, and visualize data on the squares.

We commonly use the statistical analysis tool **Statsmodel Python package** to perform linear regression. The package integrates well with Pandas and Numpy library to perform data analysis. Statsmodel has built-in tools for carrying out statistical testing and also checks how viable the set of plotting functions is.

Linear regression is one of the simplest modeling techniques, and it determines the relationship between the predictor variable (x) and its response (y). OLS operates with the assumption that the relationship takes the form of:

$$y = \beta_0 + \beta_1 * x$$

In the OLS estimator, the two β are used to minimize square distance between the values predicted and the actual values. This technique is common in linear regression since it is easy to understand the model and calculate coefficients.

Statsmodel package has several classes to modify OLS.

Example

Here, we will import data from the U.S. macroeconomic data from the website. In this case, we will load the **Longley dataset** into the Pandas DataFrame.

```
import numpy as np
import matplotlib.pyplot as plt
%matplotlib inline
import Pandas as pd
import statsmodels.api as sm
from statsmodels.iolib.summary2 import summary_col
from linearmodels.iv import IV2SLS


# extract longley dataset from the web with a column index 0
dframe =
pd.read_csv('http://vincentarelbundock.github.io/Rdatasets/csv/dat
asets/longley.csv', index_col=0)
dframe.head()# read the first five rows
```

Output

Longley Dataset from Web

GNP.deflator	GNP	Unemployed	Armed.Forces	Population	Year	Employed
83.0	234.289	235.6	159.0	107.608	1947	60.323
88.5	259.426	232.5	145.6	108.632	1948	61.122
88.2	258.054	368.2	161.6	109.773	1949	60.171
89.5	284.599	335.1	165.0	110.929	1950	61.187
96.2	328.975	209.9	309.9	112.075	1951	63.221

We can modify this example to determine the relationship between the **predictor** and the **response**.

We will use the **Gross National Product -"GNP"** as the **predictor** (*P*) and the variable **"employed"** as the **response** (*R*).

We can also add a constant term that will act as the intercept of the linear model.

```
R=df.Employed

P=df.GNP


#get the statsmodels.api library to add a constant term

P=sm.add_constant(X)


#print the output

P.head()
```

Output

Constant term,x

const	GNP
1	234.289
1	259.426
1	258.054
1	284.599
1	328.975

We can now perform regression analysis from the predictor and response by using the `sm.OLS class` and `OLS(R,P)` initialization method.

The `OLS(R,P)` function has two variables passed as array like object: *p* and *r*, where *p* can either be a Numpy array or Pandas DataFrame in form of *(n,m)*. The variable *n* will be the number of **data**

points while *m* will be the number of **predictors**. The response *r* will be either one-dimension Numpy array or the Pandas series of length *n*.

```
#call sm.OLS function (statsmodels.api library)

Estm=sm.OLS(R,P)
```

We also need to call OLS object `fit ()` method to fit our regression model. If your dataset has less than 20 samples, you will get a **Kurtosis test warning**. You may ignore the warning, as Kurtosis has a valid dataset up to 20.

```
Estm=sm.OLS(R,P)
estm = est.fit()


estm.summary()


C:/user/Faith/lib/Python3.6/dist-packages/scipy/stats/stats.py:1276: UserWarning: kurtosistest only valid for n>=20

n=16


int(n))
```

Run the code to get the results.

OLS Regression Results

Dep. Variable:	Employed	R-squared:	0.967
Model:	OLS	Adj. R-squared:	0.965
Method:	Least Squares	F-statistic:	415.1
Date:	Sat, 08 Feb 2014	Prob (F-statistic):	8.36e-12
Time:	01:28:29	Log-Likelihood:	-14.904
No. Observations:	16	AIC:	33.81
Df Residuals:	14	BIC:	35.35
Df Model:	1		

	coef	std err	t	P>|t|	[95.0% Conf. Int.]
const	51.8436	0.681	76.087	0.000	50.382 53.305
GNP	0.0348	0.002	20.374	0.000	0.031 0.038

Omnibus:	1.925	Durbin-Watson:	1.619
Prob(Omnibus):	0.382	Jarque-Bera (JB):	1.215
Skew:	0.664	Prob(JB):	0.545
Kurtosis:	2.759	Cond. No.	1.66e+03

R-Squared Linear Regression

Once you calculate the linear regression, you can determine how well the linear model fits the data using the **R-Squared model**. OLS minimizes the distance between the selected data points and the fitted line, reducing the sum of the squared residuals.

R-Squared, or the **coefficient of determination**, measures how close the program should fit the data to the regression line. That is, determine how scattered are the data points along the regression line. The program expresses it as a *percentage* (between 0 and 100%) of response variable variation results in the linear model.

It determines the relationship between the independent variables and the dependent variable. Once you come up with a linear regression model, you have to determine how well the data fits in the model. The regression model determines the difference between the observed values and the fitted values. That is, it finds the smallest sum in the squared variables in a dataset.

R-squared= Linear model variation/total variation

R-squared values are always between 0 to 100%. A 0% R-squared regression means that the model does not have any variability of the response data across its mean. The dependent variable mean is used to predict the dependent variable and the regression model. A 100% regression shows there is variability of the response data across its mean.

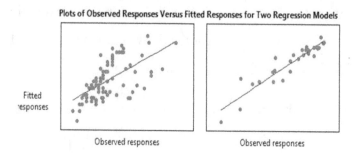

Plots of Observed Responses Versus Fitted Responses for Two Regression Models

Fitted responses

Observed responses   Observed responses

From the two plotted graphs, you will notice that the higher the R-Squared number, the better the model will fit your data. In this example, the graph on the left accounts for 38% variance, whereas the one on the right accounts for 87.4% variance. The higher the variance accounted for in the regression model, the closer the data points fitted in the regression line.

If the difference between the observed values and the predicted values are small then the regression model fits your dataset data. The fitted values should also be unbiased; they should be neither too high nor too low in the observation space.

When using R-Squared for data prediction, you may end up with a low R-Squared, indicating a good regression model or a high R-Squared value, the second which would not fit the data. Thus, R-Squared is not a good measure of showing whether a regression model is the best fit.

170

F-Statistic

F-statistic is a value returned when you do an **ANOVA test** or **regression analysis**. The test measures whether the variance between the two variables is different. The test uses f-distribution to indicate whether a group of variables are jointly significant and whether the hypothesis of the population variances are equal.

A variance is a measure of dispersion and shows how far the data scatter from the mean.

F-statistic is also referred to as f-test and it's mostly used to determine the overall linear regression value and provide the best fit to the line of regression. Compared to R-squared, f-test is more flexible and can be used in a wide variety of settings. It can also evaluate multiple data models and determine the best fit to the linear model.

F-statistic tests the equality of means and is the best variance analysis tool, especially when you want to accept or reject the **null hypothesis**. The null hypothesis is used to indicate that the model has no independent variables and fits the data well. If the independent variables in the model are not statistically significant, then the f-statistic will not be significant.

To determine whether a group of variables is equal, you can calculate the **F-statistic ratio**:

F-test= variation between sample means/variation with the samples

Chapter Summary

In this chapter, we learned how to carry out financial analysis and make the right decision when determining which financial securities to invest in. Some common financial analysis tools you should consider include:

- Calculating the daily return on each stock asset.

- Calculating the cumulative daily rate of return.

- Using moving windows to determine the volatility of a single investment.

- Calculating the number of dividends obtained from the profit proceeds of the company.

In the next chapter, you will learn about some common trading strategies you can use to maximize your profits and meet your financial needs.

Common Trading Strategies With Python

Building a Trading Strategy with Python

After you have successfully analyzed your data, the next step is to formulate a great trading strategy; something that will make you stand out from the rest and meet all your financial needs. There are some common types of trading strategies, which we will look at. You should read them carefully and understand them, so you can decide on which option will work best for you.

Common Trading Strategies

A **trading strategy** is the act of buying and selling of stock and securities on both a short-term and long-term basis, depending on their movements in the stock exchange chart. There are two main common trading strategies available: the **reversion strategy** and the **momentum strategy**.

Momentum Trading Strategy

We also know the **momentum strategy** as the **trend or divergence trading**. It is a technique in which investors rely on the recent price trend to buy and sell shares. Traders operate on the assumption that stock assets moving in a certain direction will continue to do so or follow that path until that trend loses its strength.

Entrepreneurs choose this strategy entirely based on the movement of a quantity from its current position. They believe that stocks move upwards or downwards, and with this, you can identify the direction of the momentum. The rate of a price change and trading volume affect momentum trading.

Momentum direction= Previous price- current price

There are two categories of the momentum strategy: **relative momentum** and **absolute momentum**.

Relative Momentum: This involves comparing the performance of different securities within the same asset class. The investor compares the assets against each other and buys the strong-performing security and sells the weak-performing security asset.

Absolute Momentum Trading: In this, the investor compares the current price of the security against the previous historical performance in the time series data.

Indicators of Momentum Strategy

Relative Strength Index (RSI): measures the strength of the current price trends. It shows how strong the current stock is in relation to previous stock performance. This information helps determine the time stocks are being overbought or oversold.

Rate of Change (ROC): shows the percentage change in price between the current price and the nth period of the past prices. It shows the general trend in prices and oversold and undersold securities.

Commodity Channel Index (CCI): shows the variation of the stock price based on the statistical mean. It compares asset prices against the moving average and the mean deviation of the normal price. It shows excess buying or selling when the reading is above 100 or below -100.

Moving Average Convergence/Divergence (MACD): shows the fast and slow-moving stock price average. It indicates the moving average trend lines by showing the price momentum and the price trend reversal points. If the lines are far apart, the momentum is strong; however, if they are close together, the momentum is slowing down and the price will slowly move towards reversal.

Stochastics: compares the current stock price with a certain range for over a specific period. If the range will be 20, then any reading below 20 will show an upward price momentum. If the stocks reach an overbought condition (*reading above 80*) then it will show downward price.

Examples of this strategy are inclusive of **turtle trading, moving average crossover**, and the **dual moving average crossover**.

Moving Average Crossover: occurs when the price of an asset shifts from one side of the moving average to the other. The crossover detects a change in momentum, and entrepreneurs use it as a basis for decision-making on whether to enter or exit the market.

Dual Moving Average Crossover: occurs when the short-term average changes to a long-term average. When this occurs, the momentum is shifting towards the side of the short-term average. When the short-term average rises above the long-term average, a buy signal raised. When the short-term average falls below the long-term average, a sell signal generates.

Turtle Trading: a popular strategy first introduced by Richard Dennis. The base behind this strategy is to buy stocks during their 20-day high or sell them on their 20-day low.

Reverse Trading Strategy

Reverse strategy, or **convergence trading**, involves a change in the price direction of a security. Reversal can either be *upside* or *downside*. If the price is in an uptrend direction, then the reversal will be *downside*; however, if the asset prices are on a downtrend, then the reversal will be *upside*.

We base the strategy on how the movement of quantities will reverse eventually. This occurs based on the overall direction of the price. Moving average and trend lines are major indicators in spotting reverse prices. For example, in the mean reversion strategy, we believe the stocks will return to their mean and you can deploy them once they deviate from the mean. This is practical.

A good reversal strategy should be able to show the changes in price direction either going up or going down. The strategy may occur in **intraday trading**, which happens more quickly, like within a day, a week or even within months.

Reversals are important to different investors. A long-term investor may not be interested in intraday reversal, even though they are watching the reversals on a daily or weekly basis. A day trader is more interested in price changes, and even a five-minute price reversal is very important.

Another good example to support this strategy, apart from the mean reverse strategy, is the use of **pairs trading mean reversion**, which is like the reversion strategy. In the mean reversion strategy, the stock returns to its mean. For pairs trading mean reversion, if the investor identifies two stocks to have a high correlation, the change in the price difference between these two stocks is used to signal the trading events if either of the stocks moves out of correlation with the other.

Based on this argument, if the correlation between the two stocks *decreases*, then the stock with the higher price should be in the **short position**; therefore, the investor needs to sell them. The stock with a higher price, in the long run, will return to the mean. The stock with a low price will be in a long position because its price will rise as the correlation returns to normal.

These two are the most common, but that does not mean that they are the only ones. The market has other strategies that you will come across once in a while. A good example of this is the **forecasting strategy** that tries to predict the value and direction of a stock. Another one is the **frequency trading strategy**, which makes use of the microstructure of the market.

In the next sections, we will now focus on helping you develop your own unique, valuable, and efficient trading strategy.

A Basic Trading Strategy

To formulate your trading strategy, begin with the basics of quantitative trading, which would be the moving average crossover. I will help you develop a simple trading strategy that will help maximize your investment. This strategy involves creating two different **simple moving averages (SMA)** with different periods, for example, 40 and 100 days. If you have a short moving average that exceeds the **long moving average**, then you should go for long-term investment, whereas if you have a long moving average that exceeds the short moving average, you should exit.

One thing to remember is that if you go long, then it means that your reasoning is that the stock will *rise* in price and you expect to sell at a higher price later (remember the buy signal). If you go short and sell your stock, then it means that you expect to buy it a lower price than that which you have sold it for and make a profit (sell signal).

This strategy is simple but might seem complex at first. We will, therefore, look at it step-by-step.

Step 1

Define two different time period windows: **short window** and **long window**. Afterward, set up two types of variables and assign each variable an integer data type. Make sure the integer assigned to the short-term window is shorter than the one assigned to the long-term window.

Step 2

Set up an empty signal DataFrame (*"signal"*) to help in calculating the buy and sell signals. Copy *"aapl"* index to enable you determine daily buying and selling signals of *"aapl data."*

Step 3

After this, the next step is to create a column in the empty *"signals"* DataFrame created in step two. Initialize the column and assign value [0.0] to all the rows in this column.

Step 4

Once you finished creating all the variables and assigning values to the rows of the column, create a set of short and long moving averages for both long and short time windows. Here, we will use the `rolling()` function to calculate the rolling window period.

Specify the window and the `min_period` as the arguments to the `rolling()` function. For example, we will the window to be `short_window` or `long_window`, and a `min_period` of 1, in which 1 will be the minimum number of observations in the window chosen.

After setting up the window, you need to calculate the mean average of both short and long windows. Later, you will need to calculate the rolling mean by calling the `Mean()` function.

Step 5

After getting the mean average for both short and long windows, you have to create a **signal** that shows when the short moving average bypasses the long moving average. You can do this on the condition that the window period is greater than the shortest moving average window.

```
djsignals['short_mavg'][short_window:] >
djsignals['long_mavg'][short_window:]
```

If the condition tested is *true*, then the initialized values [0.0] in the "*signal*" column will be overwritten to [1.0]. Note that from step 3, your program will only create a signal when the condition is *false* and the values remain [0.0].

We will now create a program that will enable you to generate a **real trading strategy**.

```
#import the libraries
import numpy as np
import Pandas as pd

# Initializing variables for short and long window
short_window = 40
long_window = 100

# Initialize `dfsignals` DataFrame with `signal` column
dfsignals = pd.DataFrame(index=aapl.index)
dfsignals['signal'] = 0.0

# short window over short moving average
dfsignals['short_m-avg'] = aapl['Close'].rolling(window=short_window,
min_periods=1, center=False).mean()

# long moving average over long window
dfsignals['long_m-avg'] = aapl['Close'].rolling(window=long_window,
min_periods=1, center=False).mean()

# Trade signals
dfsignals['signal'][short_window:] = np.where(dfsignals['short_m-
avg'][short_window:]
                                              > dfsignals['long_m-
avg'][short_window:], 1.0, 0.0)

# trading orders
dfSignals['positions']=dfsignals['signal'].diff()

#results
Print(dfsignals)
```

Run the program and view the signals DataFrame to understand the trading strategy.

Next, we will write a code that will plot both **short** and **long moving averages** for buy and sell signals using the Matplotlib.

```
# Import `pyplot` module as `plt`
import matplotlib.pyplot as plt

# Initialize the plot figure
figure = plt.size()

# create a subplot and label y-axis
label = figure.add_subplot(111, ylabel='Price in $')

# plot yahoo closing price
aapl['Close'].plot(ax=label, color='r', lw=2.)

# Plot moving averages
dfsignals[['short_m-avg', 'long_m-avg']].plot(ax=label, lw=2.)

# Plot buy signals
label.plot(dfsignals.loc[dfsignals.positions == 1.0].index,
           dfsignals.short_m-avg[signals.positions == 1.0],
           '^', marketsize=10, color='n')

# Plot sell signals
label.plot(dfsignals.loc[dfsignals.positions == -1.0].index,
           dfsignals.short_m-avg[signals.positions == -1.0],
           'v', marketsize=10, color='l')

# Show the plot
plt.show()
```

185

Evaluating Moving Average Crossover Strategy

Using the Pandas library, you can calculate the trading strategy metric and use the data to improve the strategy. To do this, use the **Sharpe ratio** to test whether the portfolio results are favorable and can help make smart investments.

The goal is to ensure that the investment returns are considerable and that the risks involved are manageable. When evaluating a portfolio, the greater the portfolio Sharpe ratio, the better the returns compared to the risks involved. If the ratio is greater than 1, then you can go ahead with that investment. If it's greater than 3, the investment is *excellent*.

Backtesting of Trading Strategies

Backtesting involves analysis of historical data to identify risks in any investment portfolio. It estimates the performance of any trading strategy chosen by the investor. That is, it estimates the strategy performance based on past data period.

It provides you with a step by step process in developing a trading strategy for a certain period of time. It guides you on how to approach any trading pattern and help you become a successful trader. For example, choosing a certain trading strategy may not work the way

186

you expected and as a result, you may find yourself incurring losses. To avoid this, you can backtest your trading strategy to enable you to forecast how the strategy will perform in different market conditions.

To perform backtesting on any trading strategy or investment strategy, you need to have a detailed historical data. This is very important as it provides more information on synthetic data which is otherwise unavailable. With the historical data and all trades that occurred in the past, you can use rules defined by a specific strategy to determine the success of the investment portfolio.

Initially, large institution and professionals dealing with large volumes of data did backtesting but currently almost every investor is increasingly using it. Various web-based apps have emerged making the process easier.

Most investors operate with the assumption that what worked in the past is likely to work in the future and if a strategy worked poorly in the past, it will not work well in the future.

Advantages of backtesting

· Helps to optimize the performance of trading strategy.

· It is used to develop a successful trading system.

· Helps determine market conditions that has an impact on your profitability. It also provides you with real world statistics on profitability and risks involved.

· Helps you to easily adapt to a particular stock asset

· Provides you with realistic assessment on your capability and skills to successfully carry out a particular strategy.

Disadvantages of backtesting

· It is impossible to backtest strategies that affect historic prices

· When backtesting on a strategy, the chances that what worked in the past may work in the future are minimal.

· A lot of detailed historical data is required so as to get accurate results.

You can use backtesting to determine

·   Net gain or loss in an investment. Examining historical data for a period of time using a specific strategy rules will help determine the net profit or loss in an investment portfolio. It can also determine annual returns in an investment.

·   Used to determine volatility measure: backtesting is used as a tool to measure the maximum percentile in stock prices.

·   Determine moving average: It is used to draw the percentage gain or loss in an investment portfolio.

·   Capital exposure: In capital exposure, the percentage of the total capital invested in the market is determined. Using the capital invested, a win loss ratio chart is determined.

Factors to consider when backtesting a trading strategy

1.   Always consider pricing trends from a wide market in a given time frame for the strategy being tested. If a strategy was only backtested for two years, it will not give you accurate net gains. Therefore, the best backtesting results are the ones in which a strategy is backtested for a long-time period under different market conditions.

2.   Take into account into which market the backtesting was done. If backtesting of the wide market was done on financial institutions stocks, the same strategy will not work in tech stocks. It is a general

rule that backtesting should be maintained on a particular genre of stocks.

3.   When developing the trading system, volatility measures are very important. This helps investors in making the right decision. For example, if the equity margin drops below a certain percentage, investors should keep the volatility low so as to minimize the risk.

4.   Having an increased exposure can help have higher gains in profits or higher losses. Less exposure will reduce the amount of profits or losses gained from an investment. At all times, you should keep the rate of exposure to below 70% so as to reduce risk as well as make it easier to transition into or out of any given stock asset.

5.   Using backtesting tools to determine the average gain or loss as well as determining the win-loss ratio can be essential in determining the optimal stock position. Investors can increase the average gains and win-loss ratio by taking on larger positions.

How to backtest trading strategy

Tools needed;

·   Historical data of the asset class you want to trade on

·   Trading simulation software

- Strategy definitions and rules

- Time period or number of days to perform backtesting

- A detailed analysis of the backtest on previously backtested strategy.

Backtesting can be done automatically or manually. Automated backtesting are ultimate programs designed to open, manage, and close trades for you. Automated backtesting use trading simulators or softwares which visualize historical data and provide detailed stock performance on a step-by-step basis. You can easily buy the program from the market and use it to analyze stock strategies. In manual backtesting, you have to scroll through the trading chart in the trading platform and see how the strategy will perform and areas which needs some improvement.

From the detailed information provided, you can determine whether the trading strategy will be profitable based on its performance. It also helps you to spot market conditions which can affect the performance of the strategy and how they will affect the strategy in future. From this information, you can make your decision whether a particular trading strategy is good or not.

Trading simulation software

There are various quality backtesting software online. These are the top 6 backtesting software you can use to test your trading strategy.

1.   MetaStock: This is one of the most powerful tools for traders who want a powerful backtesting and forecasting tool. It provides excellent Win/Loss report.

2.   Quantshare: This tool is suitable for those who want to develop a detailed quantitative analytical system for multiple trading strategies.

3.   TradingView: This software is great for beginners. It has inbuilt strategies and a simple sleek interface to help traders write simple code to develop a strategy.

4.   TrendSpider: Provides a cutting edge for AI, has multiple timeframes for Auto Trendline recognition and provides system backtesting.

5.   Interactive Brokers: This tool is great for IB traders. It allows traders to develop a powerful portfolio strategy and testing with no programming knowledge.

Tradestation: Provides technical backtesting procedures for all Tradestation brokers. The tool integrates with Algorithmic Trading to provide the best backtesting.

Chapter Summary

In this chapter, we learned in-depth about common basic trading strategies using Python. We also learned how to use the momentum strategy and reverse trading strategy to compare investments from different companies and make the right investment decisions.

We went over how to forecast trends for existing market prices and compare them to the current market price. Along with that, you should also be able to calculate the number of dividends received from your investment in a company.

From backtesting subtopic, you're able to learn what is backtesting and why it is important to backtest a trading strategy before investing your capital in that strategy. You're also able to know the requirements needed and how to backtest.

Conclusion

Python is one of the best high-level programming languages for carrying out data analysis, machine learning, and creating web-based applications. There are other high-level programming languages in the market, but Python remains one of the best general-purpose programming languages. Over recent years, financial institutions have increasingly been using Python for all their financial needs.

Python is a great tool for data analysis and visualization, which is an important tool for financial analysis. The Python library is rich in tools and packages for performing statistical data and mathematical calculations. You can easily integrate Python with third-party apps and build a competitive structure for your organization to address all the consumer needs.

This tutorial on basic Python in finance will provide you with the basic knowledge for relying on Python programming language and understanding your financial market. It also showed you how you can use Python to make the right investment decisions.

If you want to dive into an investment in various financial securities, this tutorial should guide you on various trading strategies and how to choose the best trading strategy. You will also learn the key principles for successful trading and how to use time series data to predict financial security trends. If you plan to do all this yourself, this book hopefully has given you the guidance for getting started and setting up your workspace to observe current market trends.

After setting up the workspace, we imported the Python Pandas library to help in data manipulation. This book provided various ways you can import data from various sources into the Pandas development framework and manipulate it. You also learned how to use Excel to manipulate data and how to integrate Excel with Python.

You can now use the *Pandas_datareader()* function to import data from an Excel file and know how to read time series data in Python. You also learned how to rely on the predicted stock trends to make your investment decisions.

When dealing with large volumes of data, you need to bring forth data analysis and visualization tools to help interpret the data presented. Data visualization is a presentation of data in an easy-to-understand format using tools like graphs, charts, and scatter matrices. Data analysts rely on these tools to present data in simple formats.

You learned how to visualize time series data using Pandas library and plot data for an easy read. You can now plot different versions of time series data and customize your charts by changing color, font size, and grid. You also learned how to use scatter matrices to plot data points and read the relationship that exists between the plotted variables.

Financial analysis information is important to shareholders as they rely on this information when decision-making. Finance statements help to determine the financial strength and weaknesses of any investment. In our financial analysis chapter, we reviewed how to use the observed data to calculate daily returns, the cumulative daily rate of returns and how to calculate dividends and volatility.

Calculating the moving average will enable you to determine whether it is worthwhile to invest in a particular financial security or not. You can calculate the trading strategy metric and use the data collected to improve the strategy. I hope you learned much from this book and can now use Python in finance with confidence.

The end... almost!

Reviews are not easy to come by.

As an independent author with a tiny marketing budget, I rely on readers, like you, to leave a short review on Amazon.

Even if it's just a sentence or two!

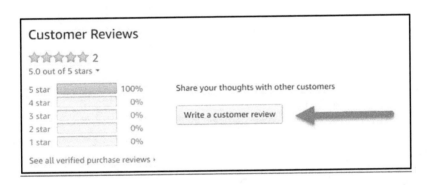

So if you enjoyed the book, please leave a review on the Amazon page.

I am very appreciative for your review as it truly makes a difference.

Thank you from the bottom of my heart for purchasing this book and reading it to the end.

References

Bowne-Anderson, H. (2018). Time series analysis tutorial with Python. *DataCamp*. Retrieved from https://www.datacamp.com/community/tutorials/time-series-analysis-tutorial

Data Analysis and Visualization with Python. (n.d.). *Geeks for Geeks*. Retrieved from https://www.geeksforgeeks.org/data-analysis-visualization-python/

Kumar, N. (n.d.). Graph plotting in Python: Set 1. *Geeks for Geeks*. Retrieved from https://www.geeksforgeeks.org/graph-plotting-in-python-set-1/

pandas.DataFrame.resample. (n.d.). *pandas 0.25.1 documentation*. Retrieved from https://pandas.pydata.org/pandas-docs/stable/reference/api/pandas.DataFrame.resample.html

Scipy Lecture Notes. (n.d.). Retrieved from http://scipy-lectures.org/

What is Momentum Trading? (n.d.). *FXCM*. Retrieved from https://www.fxcm.com/za/insights/what-is-momentum-trading/

Willems, K. (2019). Python for finance: Algorithmic trading. *DataCamp*. Retrieved from https://www.datacamp.com/community/tutorials/finance-python-trading

Time-Series with Python

Disclaimer

Introduction

The Nature of Data

Before getting into the world of time history, let's learn a little bit about data.

Two things come to our mind when we study data. The first one is that data is a representation of the real world. Consider just one example: every day; we leave our footprint on the web with every click that we make, every like that we make and every comment that we type. All these actions say a lot about our interests, and, from this, it is possible to map our consumption profiles, topics of interest, favorite artists, films and much more. By now, you must have realized that the ads that appear on the screens of our devices are related to our taste. That is why data is a perfect representation of the real world. It can retain a domain by mapping behavior from the data.

The second core concept is that within a particular characteristic that is always present or absent, data sets can be divided into two major classes:

- Timeless data
- Temporal data

Timeless data is data that does not depend on time. For example, a dataset containing images, datasets for sentiment analysis, for character recognition, or detection of false profiles, are not temporally organized datasets.

Some datasets have an attribute of 'time,' but that is not temporal data because it is not data that gets organized over time.

Temporal data, on the other hand, is organized over time, with a 'time' attribute being an index of observation in the dataset. Cool examples of temporal data are population, and social data such as HDI, Gross Domestic Product (GDP), unemployment rate, illiteracy rate, and geographic evolution. Ongoing cases of temporal data include public, meteorological, and financial data.

What is Time-Series?

Time-Series can be defined in more than one way. The first is from a book called Computational Intelligence in Time-Series Forecasting, written by Ajoy K. Palit and Dobrivoje Popovic:

"A time-series is a time-ordered sequence of observation values for a physical or financial variable made at equally spaced time intervals Δt, represented as a set of discrete values $x1, x2, x3, ...,$ etc. In practice, the sequence of values is obtained from the sensors by sampling the related continuous signals. Based on measured values and generally corrupted by noise. Time-series values usually contain a deterministic signal component and a stochastic component representing noise interference that causes statistical fluctuations around deterministic values."

That sounds complicated, and the reason it is such a detailed definition is that the book it comes from is written to address concepts and apply them in Finance and Engineering contexts.

A simpler definition of a time-series is:

"A time-series is a set of data ordered over time, in a well-defined time interval."

Much better, easier to understand.

Great! Now we need to clarify the definition.

When the second definition says, "time-series is sorted over time, in a well-defined time interval," it means that data is grouped sequentially, using two factors:

1. The recorded data of occurrence

2. The unit of time – daily, weekly, monthly, yearly, etc

The graph below shows a time-series:

On the horizontal axis (x-axis), you can see the dates the data were recorded as blue dots. The first definition refers to these blue dots as $x1$, $x2$, $x3$, ..., xn.

On the vertical axis (y-axis), you can see the attribute values for the data, in this case, $ amounts.

The last part of the first definition states that "Based on measured values and generally corrupted by noise, time-series values generally contain a deterministic signal component and a stochastic component representing noise interference that causes statistical fluctuations around the values deterministic." This has a relationship with the less trivial statistical characteristics of the time-series. These characteristics define some of the behaviors exhibited by a time-series but discussing those is outside the scope of this book.

A Brief Historical Overview of a Time Series

- Time-series theory was implemented primarily during the interwar period by Yule and Wald.

- During the 1950s, researchers at the Cowles Foundation Group developed econometric models of simultaneous or interdependent equations. These were developed to estimate the maximum likelihood methods used and were applied mostly to macro-economic models.

- During the 1960s, Holt and Winters developed the exponential smoothing technique time-series, mostly used for forecasting.

- During the 1970s, a methodology was developed to create empirical time-series models, providing a boost to applications by Box and Jenkins. This methodology states that when the information on the design is searched for, it is directed to the available data. That means it is not necessary to rely on pre-existing theories to create models.

- Box-Jenkins model forecasts were often better than large-scale macro models. However, macroeconomists found them inappropriate for a theoretical economic policy because they were empirical and thus lacked a theoretical basis.

- Until the early 1980s, there was a great deal of rivalry (see, for example, Johnston, Econometric Methods, McGraw Hill, 1986). After that came convergence.

Why is Time-Series Exciting?

No doubt, you've seen graphs showing how the stock market has evolved over the last 12 months. You've probably seen graphs showing a country's economic growth over a set period, or graphs showing the temperatures over 12 months for a country you intend to visit.

Most of you will not be aware that all of those examples are time-series because they represent numerical variables over time. In this section, we will look at what a time-series is and the important concepts that help you interpret them.

As mentioned, a time-series is a numerical variable. Time has a scale. And this scale is ordered. It can be in months, days, hours, seconds, years, decades, weeks, quarters, whatever timeframe you want to use. For example, there would be 365 values for the stock market value every day for one year. Place those values into date order, and you have a time-series.

If you opt for days, you get one value per day; if you opt to show the data by month, you get one value per month. The level of detail in your variable is dependent on the time scale you choose. Take, for example, a temperature observation for one year. You would have 12 values – one for each month. Each month would have between 28 and 31 observations, depending on how many days in the month, and that equates to 365 values per year.

The time-series resolution is equal to the number of points in a fixed period. The more points there are, the more resolution the time-series has, and that means you have more available information.

Have a look at these examples of time-series in real-life applications:

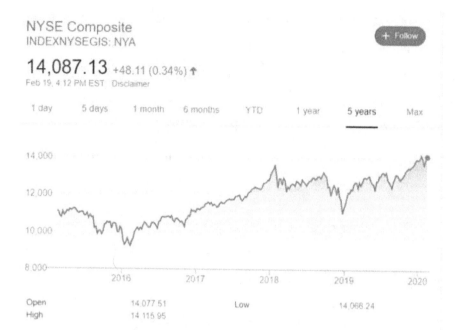

NYSE Composite
INDEXNYSEGIS: NYA

+ Follow

14,087.13 +48.11 (0.34%) ↑
Feb 19, 4 12 PM EST Disclaimer

| 1 day | 5 days | 1 month | 6 months | YTD | 1 year | 5 years | Max |

| Open | 14,077.51 | Low | 14,066.24 |
| High | 14,115.95 | | |

This is a stock market graph using the NYSE index over five years. The index shows the health of the stock market in the USA, and you can see, in the graph, the time-series trend.

Typically, mathematical techniques are used to create the trend, and, later, I will show you how this is done. For now, the trend is an excellent way to understand how your numerical variable is going up and down over time.

A country's GDP time-series is another good example. This time-series will show you the economic health of that country, and, in the example below, we can see the GDP for three major countries – the USA, Japan, and the UK – as it has evolved over the 20th century.

Growth in Real Per Capita GDP in Japan, Britain, and the US, 1870–2008
(Natural log of per capita GDP in 1990 international Geary-Khamis dollars)

The red line shows Japan's GDP, and we can see that, following WWII, it was slow to rise until the 1970s.

In simple terms, the time-series uses historical data to show the evolution and trend over time.

The example below shows the number of visits per day to a blog. The values used are the dates between July 1 and July 31, 2019, inclusive.

This graph is perfect for explaining an important point. Although the time-series is represented with a line, in reality, what you see is observations per day. The points you see in the graph are the data, corresponding to the visits per day.

So, from this example, we can calculate how many visitors there were:

- For any given day in the month

- For each week in the month

- Total visitor numbers for the month.

This data allows us to predict the potential number of visitors for the next month, for a quarter, even for the entire year.

A more complex example is the time-series below, showing the annual average temperatures between the years 1200 and 2000. This is known as a "hockey stick" graph and was responsible for no small amount of controversy between politicians and scientists.

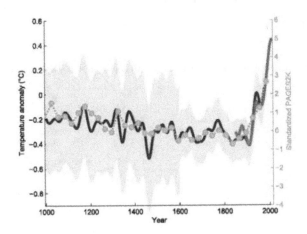

This kind of graph goes some way towards explaining the effect of greenhouse gases on global warming. The historical data was obtained from ancient trees, and the radius of the log rings used to measure the temperature.

The graph shows a light-blue shaded area; this indicates the measurement error as trees were used to measure the data and not thermometers. There is always the potential for the error to be high, but you can see that the error reduces as more historical data becomes available, and we have more accurate values to work with.

215

This particular graph shows two things – the exponential rise over thirty years to the year 2000 and, shown in red, a prediction until 2030.

Summary

You can see from these examples that we can glean a lot of information from a time-series:

1. We can understand the past and its evolution, intuiting the trend.

2. We can understand the current situation by seeing what has happened.

3. We can predict the future with a prediction error based on historical data.

These are the three main applications of the time-series.

Getting Started with the time-series and Its Evolution

In this chapter, we are going to warm up to the concepts involved in analyzing time series. We are going to look at 4 key elements that time series is used for with the relevant graphs and equations. We also look at 4 different ways to visualize time series.

Then we're going to look at more advanced time series concepts like autoregression models, moving average and ARMA/ARIMA models. While we introduce these concepts in this chapter, we investigate more applications in the subsequent.

What Elements of a Time-Series Are Interesting?

So far, you have a global picture of what a time-series is and what it can do for you. It is a numerical variable seen in time. You can see the values of the numerical variable in historical order. This helps you to see the evolution in numbers of this variable. And you can observe the trend in different periods according to the time scale you have.

Now I will show you the elements of a time-series with an example. I continue with the example of the blog visits presented above. Look again at the image above. There are two variables:

1. The first is the time scale, in this case, the days.
2. In the second column, the numerical variable (the one you want to study).

An observation of the numerical variable corresponds to each day. This is the time-series. You can draw the points of the time-series on a chart, type scatter plot (first chart). Or you can join the points with a line to visualize the historical trend of the variable.

Note that by joining the scatterplot points, you can create a graph where you visualize much better how the numerical variable goes up and down in time.

This is a time-series, and the graph is a time-series plot. But hold on to the idea that the data you have is what you see in the table. That is the point of the graph.

How close the points depend on the time scale you have.

Time-Series Properties

Depending on the area of study, books that talk about Time-series may present these properties in different ways and separate orders. So, for your better understanding, I decided to group the characteristics into two features: type and behavior.

Type

218

The type of the series is related to the structure of your dataset. Basically, time-series can be grouped into two types: the univariate series and multivariate series.

A univariate time-series contains only one attribute representing the problem domain. Just look at an example:

	Date	Temperatures
0	1981-01-01	20.7
1	1981-01-02	17.9
2	1981-01-03	18.8
3	1981-01-04	14.6
4	1981-01-05	15.8
5	1981-01-06	15.8
6	1981-01-07	15.8
7	1981-01-08	17.4
8	1981-01-09	21.8
9	1981-01-10	20.0

On the other hand, a multivariate time-series consists of several traits that represent a given problem domain. Have a look:

	Date	Location	MinTemp	MaxTemp	Rainfall	Evaporation	Sunshine	WindGustDir	WindGustSpeed	WindDir9am	...	Humidity3pm	Pressure9am	Press
0	2008-12-01	Albury	13.4	22.9	0.6	NaN	NaN	W	44.0	W	...	22.0	1007.7	
1	2008-12-02	Albury	7.4	25.1	0.0	NaN	NaN	WNW	44.0	NNW	...	25.0	1010.6	
2	2008-12-03	Albury	12.9	25.7	0.0	NaN	NaN	WSW	46.0	W	...	30.0	1007.6	
3	2008-12-04	Albury	9.2	28.0	0.0	NaN	NaN	NE	24.0	SE	...	16.0	1017.6	
4	2008-12-05	Albury	17.5	32.3	1.0	NaN	NaN	W	41.0	ENE	...	33.0	1010.8	
5	2008-12-06	Albury	14.6	29.7	0.2	NaN	NaN	WNW	56.0	W	...	23.0	1009.2	
6	2008-12-07	Albury	14.3	25.0	0.0	NaN	NaN	W	50.0	SW	...	19.0	1009.6	
7	2008-12-08	Albury	7.7	26.7	0.0	NaN	NaN	W	35.0	SSE	...	19.0	1013.4	
8	2008-12-09	Albury	9.7	31.9	0.0	NaN	NaN	NNW	80.0	SE	...	9.0	1008.9	

The Behavior of Time-Series

Time-series can present different behaviors according to some statistical properties that may be present and/or absent in the data. Let's introduce the characteristics that a Time-series can present:

- Stationary
- Linearity
- Trend
- Seasonality

Stationary

The stationarity of a Time-series is defined when all the values in the series vary around an average, and that average must be constant over time. Three elements characterize the stationarity of a Time-series: constant mean over time, constant variance, and the covariance between two observations x_t and x_{t-d} depend only on the distance between them, and that does not change over time.

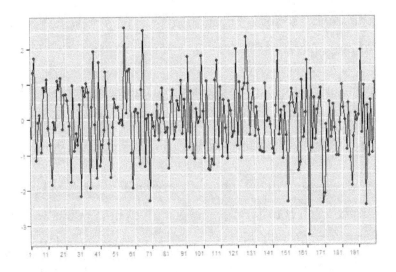

The graph of a stationary series looks like this: data varying around a constant average, which in this case is around 0 with the values between 2 and -2.

Linearity

If you already know Data Science and like it, you should know that the objective of training an algorithm is to be able to extract a model (which in the end is a mathematical function) which best represents the behavior of data to predict new values with a satisfactory degree of accuracy.

The linearity of a time-series indicates that its shape depends on its current state so that the current state can determine the model of the series. Therefore, if a series is linear, then it can be represented by linear functions of present and past values.

Since linearity allows us to check the accuracy of our measurements within the specified measuring range, linearity indicates whether the system has the same accuracy for all reference values.

For example, the manufacturer of thermometer wants to know if it provides accurate and consistent readings at five temperature levels: 210°, 208°, 206°, 204°, and 202°. Six readings are taken each time. To determine whether the thermometer is biased, individual measurements are subtracted from the reference value. The bias values for measurements with a temperature setting of 202° are calculated in the table below.

Measurement		Actual		Bias
202.7	-	202	=	0.7
202.5	-	202	=	0.5
203.2	-	202	=	1.2
203.0	-	202	=	1.0
203.1	-	202	=	1.1
203.3	-	202	=	1.3

Temperature measurements with a setup of 202° are biased positively. The thermometer indicates higher readings than the actual temperature.

For interpreting the linearity of the thermometer data, determine if the thermometer bias changes for different heat settings. If the data does not form a horizontal line in a scatter plot, there is linearity.

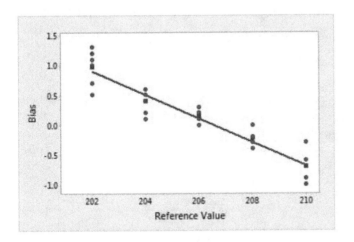

In the scatter plot, we can see that the bias changes with the increased heat setting. In lower heat settings, the temperatures are higher than the actual. In comparison, the readings of temperatures are lower than the original for the higher heat settings—the bias changes with different heat settings, which means that this data is linear.

Trend

The word trend will always be a pattern when we want to refer to something that has a given behavior and has been observed for some time; we can conclude that the trend is for that behavior to remain or change.

In Time-series, the idea of trend follows the above reasoning. When we see a graph in a time-series where the data is rising, falling, or constant and how fast these behavior changes, we see a trend. In other words, we see this behavior in the long term. The most common behaviors are the constant trend, linear trend, and quadratic trend.

Here in this graph, we can see the behavior of the indicator much more clearly. We can see that time, which is the variable under analysis here, apparently has a downward trend.

This means that, as time goes by, there is less demand for a particular product or service that is under the analysis.

Seasonality

A time-series presents some patterns of behavior. A particular pattern can be repeated at specific times over time, and that pattern is called seasonality. The word seasonality is derived from the word 'Season.'

For example: imagine a Heineken dataset for selling beer in western countries. We know that beer is a drink that is served cold and consumed in events of entertainment and the coldest seasons of the year, between December and February.

So, imagine a graph where sales have been observed for five years, where we are viewing data at annual intervals and notice that there is a peak in sales between December and February and a considerable drop in the coldest periods of the year.

Here is an example to illustrate how seasonal behavior appears in the data. Below is a graph of a seasonal series for minimum daily temperatures in Melbourne that reflects the behavior cited in the beer example:

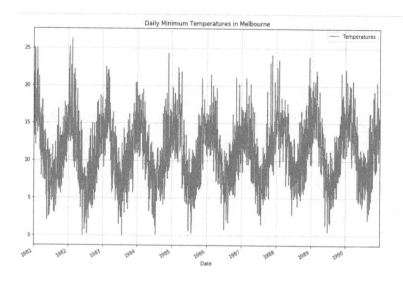

The graph plots time on an annual interval, denote the seasonal behavior. Note that the series design between one year and another has a "V" shape, which shows a peak in high temperatures between the end of one year and the beginning of the other and a drop in temperature in the middle of the year. See that the seasonal pattern is repeated over the years observed, thus showing the seasonality of the series.

The Statistical Tools to Interpret a Time-Series

The first thing is to understand the time-series as a numerical variable. You can use the same tools as if it were a single numeric variable. You can use the graphics to understand the distribution. I personally use the histogram and the boxplot

- Histogram

- Boxplot
- Q-Q plot

And calculate the numerical summary

- Half
- Standard deviation
- Median
- Interquartile Range
- Maximum
- Minimum

Histogram Example

The histogram is a graph of the distribution of a set of data. It is a special type of bar graph, in which a bar is attached to the other, that is means there is no space between the bars. Each bar represents a subset of the data. A histogram shows the accumulation or tendency, variability or dispersion, and the shape of the distribution.

A histogram is a graph appropriate to represent continuous variables, although it can also be used for discrete variables. That means, using a histogram, we can graphically show the distribution of a quantitative or a numeric variable. The data must be grouped into intervals of equal size, called classes.

Let us think that water is an indispensable element for the production of a tannery; likewise, the decision of how much water to use is decisive in the finish of the leather that is achieved.

A company is defining the ideal amount of water to use in its process, for which it performs various experiments in which it collects data associated with the amount of water used and the compliance in the finishing of the finished leathers.

The tabulated data are as follows:

Intervals (Liters)	Number of finished leathers
0.0 - 1.5	3
1.5 - 3.0	15
3.0 - 4.5	18
4.5 - 6	12
6 - 7.5	7

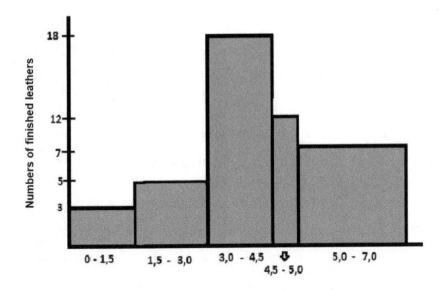

Note that the range 3.01 - 4.5 is the one with the highest number of finished leathers in good condition. In the solved example, I considered variable intervals only to illustrate a different example from those usually seen in digital and physical sources.

Example of Boxplot

Box Plot (also called box and whisker plot) is an alternative method to histogram and branch-and-leaf to represent data. The Box Plot provides information on the following characteristics of the data set: location,

- dispersion,
- asymmetry,

- tail length.

The boxplot is used to evaluate the empirical distribution of the data. The boxplot consists of the first and third quartiles and the median. The lower and upper stems extend from the lower quartile to the lowest value. The extension is not lower than the lower limit, and from the upper quartile to the highest value, it is not higher than the upper limit.

Assume that a factory produces an automotive part whose reference value is 75cm. After checking batches with out-of-specification parts, they sent two teams of workers (A and B) for training. To check the efficiency of the training, 10 pieces produced by teams A and B and 10 pieces produced by teams C and D that did not participate in the training were selected.

The data are as follows:

A		B		C		D	
75,27	74,93	74,94	74,75	75,93	73,34	75,98	76,75
75,33	74,72	75,25	74,65	76,95	74,04	75,61	76,78
74,58	74,53	75,44	74,94	75,47	75	74,2	74,74
75,01	75,32	74,62	74,92	73,6	76,18	76,44	72,58
75,71	74,05	75,35	75,46	74,85	75,33	76,84	72,86

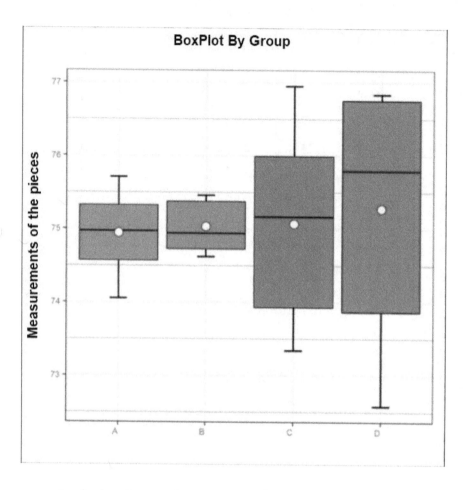

Analyzing the graph, we can see that:

1. Teams A and B produce parts with less variability, indicating that the training had the desired effect;

2. Team D is the one that produces parts with greater variability;

3. Team B is the one that produces parts with the least variability.

Considerations: As the parts of teams A and B have less variability and with an average value close to the reference value, it is worth sending the other teams for training.

Q-Q Plot Example

The QQ plot (quantile-quantile plot) is used to examine whether a certain distribution is normal or not. For this purpose, the quantiles of the empirical distribution (measured values of the sample) are compared with the quantiles of the standard distribution in a graphic. The number of possible quantiles corresponds to the number of measured values. If the data is (approximately) normally distributed, this represents the results in a diagonal line.

Assume that the height in cm is measured for a sample of 10 men: 180, 150, 160, 170, 190, 200, 210, 180, 170, and 190.

The QQ plot is to be plotted to assess whether there is a normal distribution. To do this, the empirical quantiles on the y-axis and the theoretical quantiles of the standard normal distribution on the x-axis are plotted:

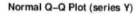

Normal Q–Q Plot (series Y)

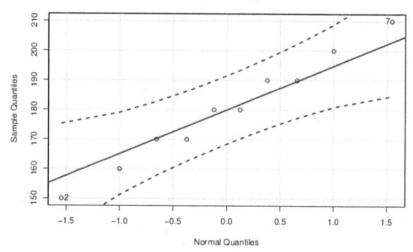

Normal Quantiles

In the plot, the approximate diagonal on which the points are located indicates a normal distribution of the data.

Now let's move on to the next section to delve deeper into the time-series.

We have already seen that a time-series is a set of data or observations that refers to one or more variables, and that is ordered chronologically. Time-series are very important in economics because almost all the variables are collected over time. That means it is interesting to see the evolution of a variable over time, not the specific value at a given moment. Hence, whenever economic variables are analyzed, economic cycles or trends are discussed.

Since the order of the data has vital importance, one must bear in mind that this modifies the analysis and interpretation of the data. Therefore, econometrics, which is responsible for finding and estimating relationships between economic variables, should take this fact into account.

Time-Series Analysis Based on its Definition and Characteristics

Given that the order of the data matters, we can say that the observations are not independent. This means that the past can affect the future. Econometrics should be aware of this characteristic and use mathematical tools that allow it to make estimations reliably. Definitely:

1. The order of the data matters.
2. The observations are not independent.
3. When estimating relationships, keep in mind that they are not independent.

Therefore, you must use different mathematical techniques and statistics.

Knowing this, then one may ask:

- What exactly does it mean that the observations are not independent?
- With what techniques are the time-series data analyzed?

Temporary Dependency

The answer to the first question refers to temporal dependence. A variable has a temporal relationship when past data affects the value of the variable in the future. For example, the long-term global GDP has a prolonged upward trend. This means that economic growth is sustained over time; therefore, what happened in the past affects the future.

On the contrary, if we roll a dice and note the date on which we roll it, we will see that there is no relationship between past and present data. In the latter case, the past does not affect the future.

Techniques for Analyzing Time-Series Data

There are many techniques to analyze time-series data. However, what is usually easier is to use a regression model. Of course, a regression model takes into account the type of time-series with which it works.

One of the most used and simplest techniques could be to modify the series or take it into account in the model. For example, eliminate the trend of a series of GDP or include a trend variable in the model. Although it is not the subject of this definition, I will give a very simple example so that you can understand it.

Let's look at the following graph:

Messi's goals in Barcelona

Evolution of the GDP of an American Country

If we calculate a regression model of the two previous series, the calculations indicate that there is a statistical relationship. However, the goals that Messi marks with the growth of an American country have nothing to do with it. However, by eliminating the trend component, it would be inferred that they have no relationship.

What is described in the previous paragraph is something that happens many times with series that are related, but when the research is done well, they do not have this problem.

A Broader Explanation of Time-Series Characteristics

Time-series analysis is an important instrument in understanding the market and in formulating action plans and strategies. The history of a variable can be used to identify periods of growth/decrease, seasonality, and also to "predict" future observations.

Time-series models are widely used to assess the behavior of a variable over time. An ice cream shop, for example, tends to sell more ice cream during the summer, while the market for coffee shops tends to be more heated in the winter.

Would it be possible to "forecast" the demand for ice cream for the summer of 2020?

In fact, the statistical models for time-series use the variable's historical past to project future observations. Thus, one can have an idea on average how the variable will behave in the coming periods.

Time-Series: Autocorrelation

The most well-known and used statistical models, such as Linear Regression and Generalized linear model (GLM), are suitable for modeling variables in which observations are independent.

The generalized linear model expands the linear model in such a way that the dependent variable is linearly linked to the coefficients and covariates with a special feedback function. This model also allows the dependent variable to have a non-standard distribution. The generalized linear model includes the most commonly used statistical models, like linear regression for responses that are normally distributed.

238

In a time-series, there is no way to disregard the dependence structure of the observations.

For example, the quantity of ice cream sold in February may be related to the quantity sold in January, which in turn may be related to that of December and so on. Thus, the use of these models can generate biased results that do not reflect the reality.

Autocorrelation is defined as an observation at a given moment that is related to past observations.

Observations can be autocorrelated in several orders. The first-order autocorrelation characterizes series where an observation is correlated with the immediately previous observation (February and January, for example). Second-order autocorrelation describes a time-series where an observation is correlated with observations at two different time units in the past (February and December, for example).

The identification of the autocorrelation is made through the Autocorrelation Function (ACF), as shown below. In addition, tests such as Durbin Watson's help to identify the first-order autocorrelation.

In the ACF example shown below, the significance of the autocorrelation order is assessed using the confidence intervals (in blue). Thus, this series has the first-order autocorrelation, since the point at lag 1 is significant.

The numbers in the graph below are correlation coefficient. So, they are each correlated with the previous observation. A correlation coefficient is a number between 1 and -1 that respectively defines a positive or negative correlation. A value of zero indicates no correlation.

Time-Series: Trend

The trend of a time-series is defined as a pattern of growth/description of the variable over a certain period of time.

There are specific tests for identifying the trend, such as the Wald Test and the Cox-Stuart Test. However, a widely used technique is the adjustment of a Simple Linear Regression to determine the slope of the trend line.

It is worth remembering that the adjustment of the Linear Regression, in this case, can lead to biased results. And to avoid this problem, robust estimators for the autocorrelation can be used. This topic will be discussed in detail in the next chapters of this book.

In the example below, we consider the data referring to the number of international passengers in air transport between the years 1949 and 1961. A clear growing trend can be seen in the series over the years.

Time-Series: Seasonality

Seasonality can be defined as patterns of behavior that are repeated at specific times of the year. For example, the number of passengers using air transport is generally higher during school vacation periods than in other months of the year, a fact that can be seen in the image presented above.

Seasonality can often be identified visually. One of the formal techniques used is the identification of significant autocorrelations of a given order. For example, if there is an annual seasonality in a series, the autocorrelation of order 12 is generally substantial.

Time-Series: Stationarity

Stationarity is an important concept in the modeling of time-series and is characterized by a variable that behaves randomly over time around a constant average.

Basically, time-series that have a trend and/or seasonality are not stationary, and it is necessary to use techniques appropriate to such a situation. In the image below, we show an example of a stationary series.

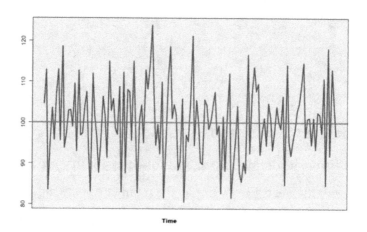

After understanding the behavior of the series and identifying the elements, it is possible to verify which model is most appropriate.

To Recap

What we have learned here:

- Nature of the data (temporal and timeless data)
- Definition of time-series
- Time-series properties and characteristics (type and behavior).

Applications of Time-Series in Making Production Forecasts

Generation production forecasts are an everyday activity of the petroleum engineer who has been developing using the implementation of computational tools based on the models' decline curve and type curves. These models have shortcomings in the accuracy of forecasts, mainly for two reasons: first, it has the excellent condition that the wells are operating in a pseudo-stable state, and second, adjust the behavior of production to a trendline, which is extrapolated in time for the prognostications.

In this section, as an alternative to these models, the use of time-series forecast generation has been proposed because they include both the trend and cyclic components and seasonal production data.

Comparing the absolute error exists between the actual data and forecasts obtained by conventional methods and the application of time-series models. Using these series yielded a better fit historical data, evidence that other trends may occur in decline (as the cube, for example) and increase the accuracy of forecasts generated.

Currently, the techniques used in generating production forecasts are based on models of "decline curve" (Arps, 1945) and "Curves Type" (Fetkovich, 1980; Palacio & Blasingame, 1993), whose use is reflected in the emergence of commercial software for tracking and forecasting production wells.

These programs integrated analysis techniques pose data, aimed at obtaining production forecasts, from generation and extrapolation of trend lines. However, analyzing these techniques can identify some points considered as disadvantageous:

- Require producing wells to be in a pseudo-stable state, that is, the pressure of flowing downhole constant. This condition is hardly satisfied due to various situations (gas accumulation in the annulus of the well, changes in the efficiency of artificial lift systems, formation damage, reduced capacity of the pipes, etc.) generate variations in the pressure conditions of the well.

244

- They assume that the variables have a parametric behavior, i.e., governed by normal trends that can be extrapolated. Actually, several things can happen here; for example, a well has several production trends throughout its life. As the phenomena mentioned above occurred, the data indicated fluctuations around the trend line, or extrapolating this trend does not allow us to obtain a good fit for the forecast.

- They have full information failures when it lacks sufficient and necessary data to implement them. This being a strong limiting field was not always updated data in areas such as fluids and petrophysical analysis of pressures.

- Do not involve the probability of occurrence of certain circumstances (such as faults in production systems, interventions wells, changing operational conditions) responses from the production system to the actions taken.

Given these reasons, the approach aims to seek new methodologies for monitoring the performance of a reservoir and predicting their future behavior, from the application of time-series.

These series are based on the assumption that current data can be calculated from the analysis of the above data, but not simply considering the tendency of the same, but include the fluctuations in variables over time, represented in both cyclical and seasonal components.

The aim is to assess how much can be improved predictions made by the time-series with respect to the results obtained with traditional models.

Decline Curve

Since the first half of the twentieth century, the oil industry has used the decline curve as the main tool for obtaining production forecasts, even though some of its considerations hardly met in practice.

A decline curve has two parameters that describe it. First, the rate of decline (D), which reflects the rate at which decreases the production of a well in terms of average daily, monthly or yearly, and is calculated using Equation 1, where q1 and q2 are respectively production rates of the well measures at times t1 and t2.

$$D = \frac{(q_1 - q_2)}{q_1 (t_2 - t_1)}$$

And in the second instance, the exponent of declination (n), that is reflected in the trend of decline curve (see Figure above), which shows that this may vary from one situation in which the decline is severe (declination is exponential with an exponent $n = 0$), to a condition of slow reduction of the production rate (harmonic decrease, with an exponent n = 1) through intermediate cases (hyperbolic decline, $0 < n < 1$).

The exponent declination can be determined in different ways, either by generating graphics, identifying trends, or using curves type.

Once we determine the declined exponent, then we can use the model proposed by Arps (https://petrowiki.org/Production_forecasting_decline_curve_analysis) (presented below) to calculate future rates of production. The produced oil accumulated in an instant determines the life of the well or the time remaining to reach a certain production rate.

$$\frac{D}{D_i} = \left(\frac{q}{q_i}\right)^n$$

The forecasts obtained with this technique are simple extrapolations of a trend line generated from the historical production data of a well and do not take into account the natural fluctuations associated with the process of extracting oil from the ground, which is why they are usually related to considerable errors. The impact of these errors is enormous since from the production forecasts, many of the investment decisions are made in the production fields, operational expenses are quantified, production commitments are generated, and oil sale contracts are generated, among many other aspects related to the management of an oil asset.

Advanced Time-Series Concepts

As already mentioned, a time-series is "a sequence of empirical data ordered in relation to time; that means, it can be plotted against time," so it is applicable to a set of data recorded periodically, whether a daily, weekly, half-yearly, monthly or annual record. Among others, some practical examples are the total annual sales of warehouses, the quarterly value of the GDP, or, in this case, the production of one or more oil wells.

To describe a time-series, the following components must be identified:

a) **Trend or trend component:** Is the result of factors affecting the data in the long term, which generate a gradual and consistent pattern of variations in the same series. In the case of a well, the trend of production data is usually the decline due to phenomena such as depressurization of the reservoir, depletion of fluids present in it, or the occurrence of formation damage. According to traditional methodologies, the trend may be exponential, hyperbolic, or harmonic. But the other trends, such as linear, quadratic, or cubic, may be considered in data analysis.

b) ***Cyclical and seasonal components:*** Both cyclical and seasonal variations can be understood as the undulating movement of data above and below the trend line. But the difference between one and the other has to do with the range of data being analyzed. In the case of cyclical components, it refers to processes that occur more spaced out within the observation interval while seasonal components occur with less separation in time.

For example, when talking about the operation of an oil well, a seasonal component may refer to the variations in the production rate that happen every day (in the hottest hours, the rate increases, while in the coldest nights, it decreases). In contrast, a cyclical component may be given by processes of gradual reduction of the flow rate generated by the damage to the formation, and a subsequent increase in the rate, after performing periodic stimulations to the well.

c) ***Irregular component:*** which relates to data that are far off-trend, such as when a production stop occurs in a well. In this case, it is necessary to make decisions regarding the handling of these abnormal values, since including them in the calculations it would add a large amount of noise to the description of the abnormal values. For this reason, a value is usually estimated to replace the anomalous data. The time-series models can be classified into four large categories, which can be identified by applying the Box-Jenkins Methodology by which the simple autocorrelation (S.A.F.) and partial autocorrelation (P.A.F.) functions are calculated. The typical forms of these two functions are presented in the Figure below.

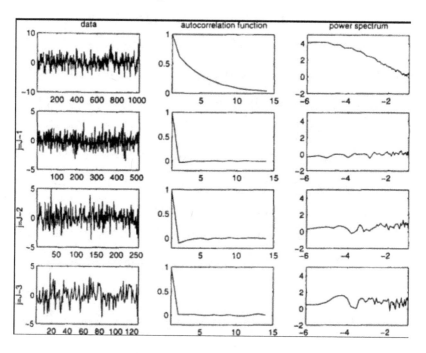

Autoregression Models

One of the characteristics that a time-series can have is that the values of the study variable (in this case, production) depend on its previous values. In mathematical terms, an autoregressive (AR) model calculates the value of the variable and at instant t using the equation shown below.

$$y_t = a + by_{t-1} + \epsilon_t$$

Where 'a' and 'b' are time-series constants, and 'ϵ_t' is the white noise assumption. y_{t-1} is the value of y at instant t-1. In this case, an AR model of order one is being referred to, because when calculating the value of the variable of interest (y) at instant t, is taking a previous data to which you want to predict (in this case y_{t-1}). Similarly, there are expressions for higher-order models, including a higher number of adjustment constants.

Additionally, the determination of the order of the model (i.e., the amount of previous values that should be used for the calculation of future value) is done by determining the "P-value" that is considered a measurement factor of how likely the model's predictions will come close to reality. This parameter ranges from zero to one, and the higher the "P-value," the more likely the proposed model will describe the data accurately.

Moving Average Models

In some cases, the data of a time-series depends on the white noise that is associated with each of the observations that make up the series. In this case, a white noises error term is added (ε_t). In addition, a second order white noise term is added that is related using a paramer (θ_i). The general form of a moving average model is presented in the equation below.

$$y_t = a + \epsilon_t + \theta_1 \epsilon_{t-1} + \ldots + \theta_q \epsilon_{t-q}$$

Where a is the mean of the series, θ_1, θ_2, θ_q, are parameters to be determined for the model, and q is the model order, i.e., the number of past values affecting the calculation of the current value. ε_t is a white noise error term at time t

If the model order moving average is 1, their expression would be presented in the equation below.

$$y_t = a + \epsilon_t + \theta_1 \epsilon_{t-1}$$

"ARMA" and "ARIMA" Models

The analysis of RA and MA models are simple, as one is simply considering a component of the time-series (the trend of the latest data or the error of the latest data, respectively). Still, it is unusual for this to happen, as in most cases, both influences coexist.

In these situations, we must resort to autoregressive models of the moving average, better known as ARMA(p,q) models, where p and q respectfully represent the autoregression order and the moving average order. The general expression of an ARMA(1,1) model would be the same as presented in the below equation.

$$y_t = a + by_{t-1} + \epsilon_t + \theta\epsilon_{t-1}$$

The models presented (RA, MA, and ARMA) have a common characteristic: they are used to represent stationary stochastic processes. That means the variables are ordered chronologically, and the probability distribution of each observation remains constant over time.

However, in some cases, this condition of a stationary time-series is not met, and making forecasts using models such as those already seen will have some associated error. For this reason, Box and Jenkins proposed ARIMA models (https://www.machinelearningplus.com/time-series/arima-model-time-series-forecasting-python/), which, in addition to containing an autoregressive and a moving average component, include an integration component. Hence its name stands for **"autoregressive integrated moving average."**

This term of integration numerically intended to lead a non-stationary series to have a stationary behavior. In this case, the numerical models are more complex, as presented in the following equation.

$$y_t = a + b_1 y_{t-1} + ... + b_q y_{t-q} + \epsilon_t$$
$$+ \theta_1 \epsilon_{t-1} + ... + \theta_q \epsilon_{t-q}$$

Precisely because of the large amount of data to be analyzed, the complexity of the calculations and the huge number of options for orders of the models is concerned in the process, it is recommended that this type of statistical analysis must be carried out by personnel with great experience in handling data with the help of specialized software.

Methodology Analysis of Production by Time-Series

To assess the accuracy of the predictions obtained using time-series analysis of production data of three wells located in the basin of the Eastern Plains was performed using the discussed methodology and comparing these results with which they were obtained employing four commercial programs:

- Oilfield Manager (OF M)
- ValNav
- Topaze
- Ryder Scott Forecast

The method used to predict the production employing time-series is summarized in four steps:

1. Handling of irregular data,
2. Historical trend adjustment,
3. Preparing the time-series model
4. And forecasts generation.

1. Handling of Irregular Data

Including the zero production values in the analysis of the data is not convenient. Let's assume that a failure had occurred that prevented the production of a certain volume of oil on a specific day. If this failure had not occurred, the production of the well would not have been zero. So, it's not accurate to include it in the oil production capacity calculations. Therefore, it is recommended to assign some value to the date in which the variable data appear. In this case, the "moving average" is used, a technique in which the missing value is replaced by the average of the data from the days before the date in question. In this case, data from the week prior to the event were used, according to the following equation.

$$q_i = \frac{q_{i-7} + q_{i-6} + ... + q_{i-2} + q_{i-1}}{7}$$

This assumption is based on a statistical principle, according to which, for a data group with normal distribution, the expectation is equivalent to the average.

2. Historical Trend Adjustment

Regression analyses are statistical processes that seek to establish the relationship between variables. If we have measurements of a certain variable (for example, the production of a well) over time, we will try to establish the behavior of that variable with respect to time using an equation.

In this case, the application of the least squares method proceeds to verify the adjustment of production data to different trends: linear, quadratic, cubic, and exponential.

The below table presents a summary of the information criteria that allow the adjustment of the data to the proposed trends to be verified:

Model	Linear	Quadratic	Cubic	Exponential
R^2 Adjusted	0.7086	0.85	0.86	0.86
MSE	$4,2 \times 10^6$	$2,2 \times 10^6$	$2,0 \times 10^6$	$2,14 \times 10^6$
AIC	15.25050	14.61050	14.52807	14.54678
BIC	15.52311	14.61441	14.53328	14.54938

- **The determination coefficient (R2):** It represents the relationship between the variations (deviations) explained by the model and the existing total variations. That means a value of R2 equal to one means that the model describes all the values that present some variation with respect to the trend. As a decision criterion, the model that shows the R2 value closest to the unit must be selected.

- **The MSE:** This criterion, as its name indicates, measures the average of the errors of each data squared. Therefore, the smaller the MSE value, the better the model studied.

- **The Akaike information criterion (AIC):** This is a measure of the relative quality of the model being considered for the description of the phenomenon. It presents an estimate of the information lost when the model being considered is used to describe the data. In other words, the smaller the value of the AIC, the less information is being lost, and therefore, the model is more accurate.

- **The Bayes Information Criterion (BIC):** Analogous to the AIC, this criterion allows the best model to be identified. The BIC relates to the deviation of the real data from the model, including some Bayesian corrections. The definition says that "BIC includes the unexplained variation of the dependent variable," which indicates the weaker the description of the variable, the higher the value of the BIC. In other words, the best model is the one with the lowest value of the BIC.

As we can see in this case, the best adjustment was presented for the cubic trend, to the detriment of the exponential adjustment that is commonly used.

The Figure below shows the historical production observations of one of the study wells, along with the proposed trend curves.

3. Preparing the Time-Series Model

Two very important aspects must be determined at this point, the type of model and the order of the model.

The first aspect is related to the identification of an equation that more precisely allows describing the historical production data, including its trend and its cyclical and stationary components. To this end, the BoxJenkins methodology must be applied, with which the "Simple Autocorrelation Function" (S.A.F.) and the "Partial Autocorrelation Function" (P.A.F.) of the data are generated. The Figure below shows the graphs obtained numerically for the data of the study well, verifying that an ARIMA model can be used in the subsequent analysis since the graphs coincide with that proposed in the figure of *Irregular Components* part of this book.

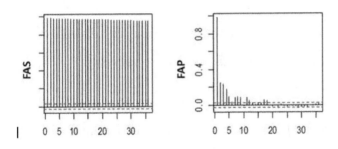

Then, through trial and error, it is necessary to determine the order of the model, i.e., how much backward data should be used to calculate a new value in the time-series.

This analysis is performed by calculating the "P Value," which relates to the adjustment of the model, as mentioned earlier. As shown in the table below, in this case, the selected model is an ARIMA (10, 1, and 10) because it has the highest p-value of 0.9965.

Order Model	1,1,1	2,1,2	3.1.3	4,1,4	5,1,5	6,1,6
Value P	$8{,}9 \times 10^{-10}$	$5{,}07 \times 10^{-9}$	$4{,}01 \times 10^{-9}$	0.01109	0.05985	.4054
Order Model	7,1,7	8,1,8	9,1,9	10,1,10	11,1,11	12,1,12
Value P	0.07178	0.9935	.9943	0.9965	.8080	.6954

4. Forecasts Generation

This is the final stage of the process where calculations are made based on the obtained model to establish future values.

Due to the order of the model, the cyclic and stationary components affect the forecasts in the short term (as can be seen in the figure below). Still, this effect is reduced when it is intended to continue advancing in time, getting to obtain a trend line after two months of forecasting simply. In this case, the production was predicted from day 5006 of operation of the well. It should be noted that between days 5045 and 5170, the production stops were recorded that the model developed is not able to predict.

The next figure shows a comparison of the results obtained for the four commercial software and the forecasts generated with the time-series model.

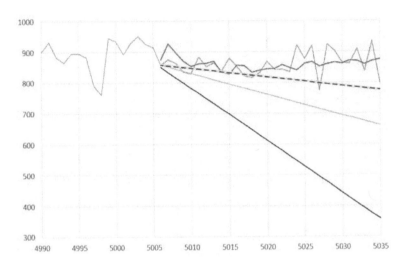

Results

The table below shows a summary of the errors obtained after generating production forecasts by five methods for a period of one month for three different wells. Since some programs use similar algorithms, the results obtained from them can be very similar.

Technique	Absolute error (%)		
WATER WELL 1 2 3			
OFM	5.34	7.18	1.75
Topaze	28.58	15.51	2.05
ValNav	11.64	9.41	2,14
Forecast	5.39	8.31	1.71
Time-series	3,51	3.30	2,14

In this case, the data generated by the Oilfield Manager (OFM) and Forecast programs have very close values, which is why they are superimposed lines graphically, and the error percentages of both programs are practically the same.

As it can be seen, the implementation of the time-series methodology allows for a significant reduction in prediction error, thanks to the fact that these models consider both the production trend of the well (which is also implicit in the software tools) and the oscillations that naturally occur in the well.

The analysis was repeated for a more extended period (one year), obtaining the results presented in the next Table. The accuracy of the results is inferior, because in a period as long as one year, the small initial deviations of the models become large margins of error.

Technique	Absolute error (%)		
WATER WELL	1	2	3
OFM	75.20	55.77	46.62
Topaze	257.78	62.37	51.28
ValNav	61.73	72.40	60.71
Forecast	73.85	51.23	44.36
Time-series	25.21	22.86	18,15

Although the behavior of the time-series model approaches a trend line after a few observations, there is less error compared to the results of the different programs.

From the analysis of the production data, it can be seen that the exponential adjustment is not always the one that allows the best description of the data behavior.

Although production forecasts generated by a time-series model reduce their effectiveness in the long term, it continues to be a tool with greater precision than traditional methodologies.

It is necessary to define a calculation algorithm for each well since the generalization of historical analysis and prediction techniques leads to the occurrence of significant errors.

Basics Concepts of Time-Series Analysis With Python

This chapter is all about getting started with Python in Time Series. We start off with why Python is the best software for analyzing time series. Then, we go through a step-by-step procedure to install Python on your computer.

Then we go through how to read and handle Time Series data into Python.

Why Use Python for Time-Series Data?

Python is an open and free general and dynamic purpose programming language. It is considered the easiest language to be read and learned. It has wide versatility, which is why it can be used in both the academic and business fields. It has object-oriented programming and a powerful computational capacity through the use of algorithms written in Fortran or in C. It also has a huge catalog of modules for all kinds of purposes: database management, graphics, artificial intelligence, etc. In recent years it has stood out for its use in the field of data analysis.

The reasons the data scientists think are worth highlighting Python as the best program for Time-series analysis are:

264

Python is Great for Beginners

First, let's answer a simple question: what makes a programming language "easy to learn"?

In my view, two factors are essential:

- Simplicity
- Learning speed

Simplicity

Python has several features that make it a so-called simple language.

First, it is a high-level language, that is, it is not necessary (but advisable) to know hardware details, or low-level languages to start learning Python.

Requires less code to complete basic tasks (such as loops, decision structures, exception handling, class declaration, etc.) when compared to other languages, such as Java (code 3 to 5 times smaller) and C (5 to 10 times smaller). Smaller code = less chance of error!

Easy Learning Curve

Because it is a simple programming language, anyone who starts programming in Python will quickly be developing sophisticated and robust programs!

By learning the basics of Python, you will be able to develop:

- Web systems with Django, Flask, Pyramid, among other web frameworks;
- Multiplatform applications with Kivy (learn more about Kivy by clicking here);
  - Link: kivy.org/
- PyGame games;
- Graphical interfaces with Tkinter or PyQT;

Versatility

Python has more than 130,000 third party libraries (exactly 135042 - at the time of writing)!

These libraries make Python very useful for specific purposes, from traditional web development, or image processing, to cutting edge technology, such as artificial intelligence, Machine Learning, and Deep Learning.

The language that helps a biomedical doctor to study the genetic sequencing of a species can also be the language that allows a data scientist to identify fraud and money laundering attempts worldwide!

Multiplatform

Because Python is an interpreted language, not compiled for a machine language, it can run on different platforms.

But first, let's understand the difference "interpreted vs. compiled"!

An interpreted language is one in which its source code is read by an interpreter and converted into executable code, which will be executed by a virtual machine. In Python, this process is divided into four steps:

- *Lexical analysis:* Breaking the source code into tokens (strings with identified meaning).
- *Parsing:* Process of taking these tokens and generating structures that show the relationship between each symbol (in the case of Python, this structure is an Abstract Syntax Tree - AST).
- *Compilation:* Transformation of an AST into one or more Code Objects ("pieces" of executable code).
- *Interpretation:* In possession of the Code Objects, the Interpreter then executes the codes represented there.

A compiled language is processed by a compiler directly from the source code for machine language specific to a processor and operating system.

Therefore, Python (which is an interpreted language) runs on different platforms, as it is enough to have an interpreter to process the source code on any system or processor. And because of this, there is no language like Python that is so intuitive for Time-series analysis for scientific or other types of research work that requires a massive amount of data visualization.

We shall start off with the basics of installing Python on your computer system.

In addition, Python is a modern programming language that can be legally obtained from the Internet for free. This is perhaps one of its main advantages it offers since scientists and their students can easily carry out their research without worrying about situations such as the acquisition of software licenses. Furthermore, the Python language lends itself naturally to applications in science, since several teams have contributed open-source libraries in various fields of knowledge, which extend the capabilities of this programming language.

How to Install Python On Your Windows / Linux / MacOS System

Python is an interpreted programming language with a multi-paradigm which supports object-oriented programming, imperative programming, and functional programming. Because of its simple and very readable syntax, it is a perfect option for those who want to learn to program. And since it comes with multi-paradigm and multiplatform features along with many modules and functions, it is also an excellent option for experienced developers too.

Whether you are a beginner or an expert, to use Python on your system, you need it installed on your computer. Therefore, you must follow the below steps to install (or update) Python as per the operating system of your computer.

Windows

Microsoft Windows is the most commercial and most commonly used desktop operating system in the world. Generally, when you buy a new computer, it is a default version of Windows. Therefore, the different ways of installing Python on these systems will be detailed below.

Python.org is the official site of the programming language. There, you will find all the information regarding the language, its documentation, and the links to download the installers of the different versions of Python that are found with the "Downloads" button.

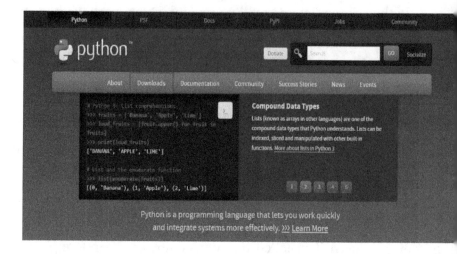

If you point the cursor on the Downloads button, a more detailed menu on possible download options will be displayed.

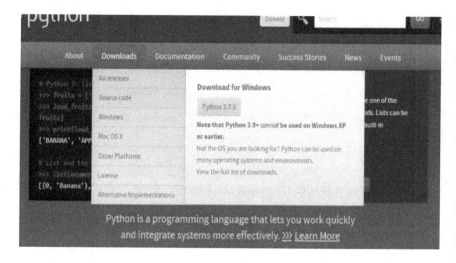

You can download the installer by pressing the "Python 3.7.3" button or, in case you need to download another version, you can press the "Windows" button. When doing so, it loads a screen similar to the following.

Python Releases for Windows

- Latest Python 3 Release - Python 3.7.3
- Latest Python 2 Release - Python 2.7.16

Stable Releases

- Python 3.6.9 - July 2, 2019
  Note that Python 3.6.9 cannot be used on Windows XP or earlier.

  - No files for this release.
- Python 3.7.3 - March 25, 2019
  Note that Python 3.7.3 cannot be used on Windows XP or earlier.

  - Download Windows help file
  - Download Windows x86-64 embeddable zip file
  - Download Windows x86-64 executable installer
  - Download Windows x86-64 web-based installer

Pre-releases

- Python 3.6.9rc1 - June 18, 2019
  - No files for this release.
- Python 3.7.4rc1 - June 18, 2019
  - Download Windows help file
  - Download Windows x86-64 embeddable zip file
  - Download Windows x86-64 executable installer
  - Download Windows x86-64 web-based installer
  - Download Windows x86 embeddable zip file
  - Download Windows x86 executable installer
  - Download Windows x86 web-based installer

Choose the desired Python version from this page, here I am downloading the version for 32 bit so be sure about the bit of your OS to get the same version of python. Once the package has been downloaded, you can install it by double-clicking the installer, and it will open a screen similar to the following:

Check that the box that says, "Add Python 3.7/your downloaded version to the PATH" is active and then press "Install Now." It will load a screen with a progress bar that will indicate each step that the installation process does.

Once the progress bar is full, a screen with options to start a Python tutorial will appear here, consult the documentation or see the changes that are made with respect to previous versions in the version that was installed is displayed. Then press the "Close" button to complete the installation process in the system.

Microsoft Store: Python can also be installed from the Microsoft Store. "Python" is added in the Microsoft store's search engine; from there, select Python.

Once loaded, press the "Get" button to get started with the download process. When the download is finished, Python will be installed on Windows.

Linux

Linux distributions are the operating systems preferred by programmers and software developers as it is the most suitable environment for their activity. These systems have a highly versatile command interpreter, and adaptable to the needs of each user as well as have highly customizable graphic interfaces in the same way.

Fortunately, Linux distributions come with Python installed by default, but sometimes the version of Python that is installed on the system is not what the user wants or needs. Therefore, shortly I will discuss the steps to follow to install the required Python version below in case it is not installed on the system.

The first step is to access the official Python site, Python.org, where the files to be downloaded will be found. Similar to the Windows process, you need to position the mouse cursor on the "Downloads" button after that a menu will be displayed containing options to choose the system in which Python will be installed.

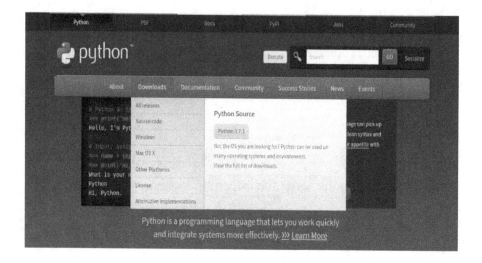

If the version that is needed is the one shown on the right side of the menu, press the button that says, "Python 3.XX", in this case, it is "Python 3.7.3". If you require a different version, press the "Source code" button. In this section, it will load a screen that contains the links to download the different versions of Python that exist.

Python Source Releases

- Latest Python 3 Release - Python 3.7.3
- Latest Python 2 Release - Python 2.7.16

Stable Releases

- Python 3.6.9 - July 2, 2019
  - Download Gzipped source tarball
  - Download XZ compressed source tarball
- Python 3.7.3 - March 25, 2019
  - Download Gzipped source tarball
  - Download XZ compressed source tarball
- Python 3.4.10 - March 18, 2019
  - Download Gzipped source tarball
  - Download XZ compressed source tarball
- Python 3.5.7 - March 18, 2019
  - Download Gzipped source tarball

Pre-releases

- Python 3.6.9rc1 - June 18, 2019
  - Download Gzipped source tarball
  - Download XZ compressed source tarball
- Python 3.7.4rc1 - June 18, 2019
  - Download Gzipped source tarball
  - Download XZ compressed source tarball
- Python 3.8.0b1 - June 4, 2019
  - Download Gzipped source tarball
  - Download XZ compressed source tarball
- Python 3.8.0a4 - May 6, 2019
  - Download Gzipped source tarball

Once you locate the desired version, the download process of the file will start with the link "Gzipped source tarball." After the file is downloaded, the content needs to be extracted. The command interpreter opens, and with the cd command, you have to change to the extracted directory.

cd / directory / Python-XXX

Once in the directory, you have to execute the installer configuration file.

./configure

Then, you should compile the file with the make command, and test the system compatibility with make test.

make

make test

After this, it is necessary to know how the Python version will be installed. It is possible to define the version to be installed as the main version of the system, and it is possible to install it as an alternative version. It is not recommended to install it as the main version since problems in the system may be caused by compatibility problems between the different versions of Python. In case you want to install as the main version, you will need to execute the following command:

sudo make install

If the version is to be installed as an alternative version (recommended), then you have to execute the following command:

sudo make altinstall

In this way, Python will be installed on the system. It should be noted that depending on the version of Python that is installed (if it is installed as an alternative version), the way to start the Python interpreter will change. If the system has Python 3.7 by default, and Python 3.6 is installed as an alternative, to start the Python 3.6 interpreter, the command needs to be executed is:

python3.6

The way to start Python 3.7 will remain the same.

macOS

macOS is the operating system of the personal computers of Apple Inc. It is widely used throughout the world because of the great efficiency it has. Installing Python on these systems, as in Linux, is generally not necessary, as it comes with pre-installed version 2.7. However, a version different from the one installed by default in the system can be installed. To do this, there are different paths, among which one is the installation process detailed above for Linux. Here are the details in the simplest way that does not require extra software to install Python on macOS:

At the same facilities in Linux and Windows, the process begins by accessing the section downloads of the (macOS) operating system of Python.org and choosing the desired version.

Python Releases for Mac OS X

- Latest Python 3 Release - Python 3.7.3
- Latest Python 2 Release - Python 2.7.16

Stable Releases

- Python 3.6.9 - July 2, 2019
  - No files for this release.
- Python 3.7.3 - March 25, 2019
  - Download macOS 64-bit/32-bit installer
  - Download macOS 64-bit installer
- Python 3.4.10 - March 18, 2019
  - No files for this release.
- Python 3.5.7 - March 18, 2019
  - No files for this release.
- Python 2.7.16 - March 4, 2019
  - Download macOS 64-bit/32-bit installer
  - Download macOS 64-bit installer
- Python 3.7.2 - Dec. 24, 2018
  - Download macOS 64-bit/32-bit installer

Pre-releases

- Python 3.6.9rc1 - June 18, 2019
  - No files for this release.
- Python 3.7.4rc1 - June 18, 2019
  - Download macOS 64-bit/32-bit installer
  - Download macOS 64-bit installer
- Python 3.8.0b1 - June 4, 2019
  - Download macOS 64-bit installer
- Python 3.8.0a4 - May 6, 2019
  - Download macOS 64-bit installer
- Python 3.8.0a3 - March 25, 2019
  - Download macOS 64-bit installer
- Python 3.7.3rc1 - March 12, 2019
  - Download macOS 64-bit/32-bit installer
  - Download macOS 64-bit installer

In this case, Python 3.6.0 will be installed with the automatic installer. The process is started by executing the downloaded file. A screen with information about the installation package will start. Proceed by pressing the "Continue" button.

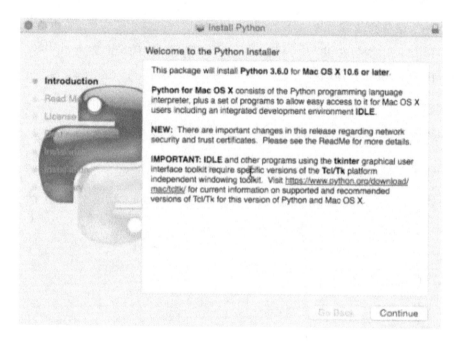

On the next screen, you will find the "Read Me" and the License information. In both steps, accept the license agreement by pressing "Agree" and then press the "Continue" button.

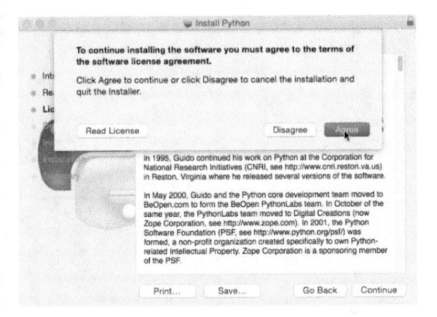

After this, information appears regarding the type of installation that will be done, and the weight of the software here clicks on the "Install" button.

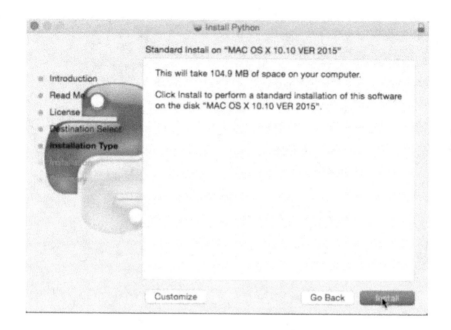

After entering the user credentials, the installation process will begin.

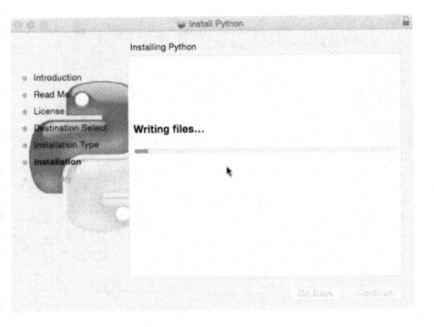

Upon reaching 100%, the installation will be finished, and Python will already be installed on the system.

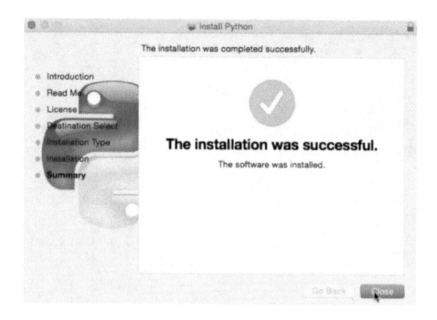

After following some of these ways to install Python, depending on your operating system, you will have the right environment to start analyzing your Time-series or other projects in Python.

Now that we have Python installed and working, let's get going. Let's start off with the reading of time-series data using Python and calculation of statistics (graphing and autocorrelation). I will explain the process of automatic SARIMA model estimation. After reading this, you will learn how to implement the simple Box-Jenkins method in Python.

Time-series analysis is, as the name implies, a method of analyzing time-series data. Time-series data is the data that increases daily (or weekly, monthly, yearly), such as "daily sales data," "daily temperature data," and "monthly airplane passenger count." Time-series data often has a relationship, such as "Yesterday's sales are similar to today's sales."

So, if you use time-series data well, you might be able to predict future sales data from yesterday's sales data. By learning time-series analysis, you can create models that predict the future from the past.

Here, we will explain how to use time-series analysis with popular Python and how to predict the future through modeling. This section deals with the Box-Jenkins method, mainly using the ARIMA model.

Reading Time-Series Data

First, we are going to load the required libraries at once, as shown in the image below.

```
# Load basic library
import numpy as np
import pandas as pd
from scipy import stats

# Draw graph
from matplotlib import pylab as plt
import seaborn as sns
% matplotlib inline
sns.set ()

# Make the graph landscape
from matplotlib.pylab import rcParams
rcParams [ figure.figsize ] = 15, 6

# Statistical model
import statsmodels.api as sm
```

This time, we'll cover data on the number of airline passengers per month. The period is from January 1949 to December 1960.

The first appearance is "Box, GEP, Jenkins, GM and Reinsel, GC (1976) Time-series analysis, Forecasting and Control, Third Edition, Holden-Day, Series G." You can check the details of the data here.

Data Link: https://stat.ethz.ch/R-manual/R-devel/library/datasets/html/AirPassengers.html

Also, the data can be found as a CSV file from this site, and here you can download the file.

CSV Link:

https://www.analyticsvidhya.com/blog/2016/02/time-series-forecasting-codes-python/

Download Link:

https://www.analyticsvidhya.com/wp-content/uploads/2016/02/AirPassengers.csv

The first column of this data is the date. Please note that if you open and save it in Excel, it will be converted to a format you would not understand well. To open the file, use a text editor such as Notepad.

Now let's load the data first in the normal reading mode.

```
# Read data normally
# https://stat.ethz.ch/R-manual/R-devel/library/datasets/html/AirPassengers.html
dataNormal = pd.read_csv ('AirPassengers.csv')
dataNormal.head ()
```

You can read the data, but in this way, the first column is treated as "data." The first column is just a date. So, it is necessary to specify and read, "The first column is a date index." In addition, the SARIMA model will be estimated later, but at this time, if the data is int type (integer), an error will occur, as follows.

```
# Read in date format (If you do not read with dtype = float, an error will occur later when estimating the ARIMA model)
dateparse = lambda dates: pd.datetime.strptime (dates, '%Y-%m')
data = pd.read_csv ('AirPassengers.csv', index_col = 'Month', date_parser = dateparse, dtype = 'float')
data.head ()
```

If the date format is simple, it is easier to read as follows.

```
# The following code is easier to read
data = pd.read_csv ('AirPassengers.csv',
                     index_col = 'Month',
                     parse_dates = True,
                     dtype = 'float')
data.head ()
```

As you can see from the JupyterNotebook results, the first column is just a date index. The data is a monthly basis with the number of days added to it (all are assumed to be the first day). This makes it easier to treat as a date.

Since the data is only one column, there is no need to keep it in the data frame. Extract only the series of passenger numbers and store it in a variable named "ts." From now on, we will proceed with the analysis using this variable called "ts."

```
# Use date format
ts = data ['# Passengers']
ts.head ()
```

286

Handling of Time-Series Data

First, plot the data. If you do not draw a graph, data analysis will not begin. It can be plotted with the following one line.

plt.plot(ts)

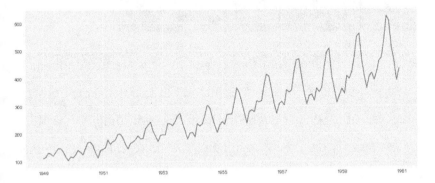

When you draw a graph, you can understand many things by yourself from it. First, the number of passengers is increasing year by year. Then, the number of passengers is likely to change periodically in each season. I will model the presence or absence of seasonal variation later. First, let's get used to handling time-series data.

To get data for a specific year and month, you can try the following:

287

```
# Data acquisition method # 1
ts ['1949-01-01']

# Data acquisition method # 2
from datetime import datetime
ts [datetime (1949,1, 1)]

# Get all the data of 1949
ts ['1949']
```

The methods for acquiring three types of data are listed together. The first and second have the same result, and in this instance, only the data for January of 1949 is obtained. The interesting thing is the third method, if you specify only the year, all data for the year, that is, monthly data will get 12 data.

The next step is how to shift the data and take the difference.

Literally, shifting means "shifting the data," and by using the shifting, we can quickly obtain the difference between the data. For example, if you want to find out how many passengers increased in February of 1949 compared to January 1949, you can calculate the difference.

In the case of time-series analysis, logarithmic differences are often taken. The logarithmic difference is an index that roughly represents the "rate of change." Another advantage is that taking the logarithm makes

288

the data easier to fit the model. This time, we do not use the logarithmic difference series, but just check the calculation method.

Here is a summary of how to calculate shift, difference, and logarithmic difference.

Shift applies the "shift ()" function: ts.shift ()

If you take out only your head, it looks like this: ts.shift ().head ()

Here is the result:

```
Month
1949-01-01 NaN
1949-02-01 112.0
1949-03-01 118.0
1949-04-01 132.0
1949-05-01 129.0
Name: Passengers, dtype: float64
```

The first data point is NaN because there is no value prior to it. The second data point (112.0) was the first point. The third (118.0) was the second, and so on.

Now, the difference between the first data set and second data set is calculated to get the actual change:

diff = ts – ts.shift ()

diff.head ()

```
Month
1949-01-01 NaN
1949-02-01 6.0
1949-03-01 14.0
1949-04-01 -3.0
1949-05-01 -8.0
Name: Passengers, dtype: float64
```

The logarithmic difference is simply taking the logarithm and then subtracting.

logDiff = np.log (ts) – np.log (ts.shift ())

If you are worried about "NaN" appearing in the result, apply the "dropna" function as follows.

logDiff.dropna ()

```
Month
1949-02-01    0.052186
1949-03-01    0.112117
1949-04-01   -0.022990
1949-05-01   -0.064022
1949-06-01    0.109484
Name: #Passengers, dtype: float64
```

Autocorrelation Coefficient Estimation

"Autocorrelation" shows how similar the previous term and the current term are. If you have a positive autocorrelation, you know that if there were more passengers last month, there would be more this month. The opposite is true for negative autocorrelation.

Autocorrelation in the statistics means that the values / measured values of a variable are correlated with themselves. This is regularly the case if the measured values of a variable are recorded or measured over time.

For example, a newspaper publisher records the number of subscribers every month. In January, for example, there were 100,000 subscribers. In February, there were 102,000. There is a connection between the individual results: a large proportion of the subscribers counted in February were already available in January (some have probably canceled; others have been gained). The number of subscribers in February, therefore, depends on the number in January, and it is the same in the other months (the number of subscribers in March depends on that in February, etc.).

To understand better, assume that the unemployment statistics show 3.56 million unemployed in March, in April it is 3.48 million and in May 3.42 million. These values are related. A large number of the unemployed from March are still without work in April and also in May

- there are not suddenly 3, x million other people unemployed. Therefore, the number of unemployed in a month is always related to the number of the previous month - there is autocorrelation.

A counterexample is a random number in roulette. Whether the 36, 17, or 4 fell earlier does not affect the subsequent run of the ball. If you look at the fallen roulette numbers over night as a diagram, you cannot draw a wavy curve like with the unemployment statistics. Instead, the view falls on a jagged curve that alternates wildly between the numbers 0 to 36.

So, let's look at an autocorrelation example here. We are going to look at the airline passenger data from the previous example and do a autocorrelation analysis in the code below:

```
# Find autocorrelation
ts_acf = sm.tsa.stattools.acf (ts, nlags = 40)
ts_acf

# Partial autocorrelation
ts_pacf = sm.tsa.stattools.pacf (ts, nlags = 40, method = 'ols')
ts_pacf
```

The results are long:

```
array([ 1.        ,  0.95893198, -0.32983096,  0.2018249 ,  0.14500798,
        0.25848232, -0.02690283,  0.20433019,  0.15607896,  0.56860841,
        0.29256358,  0.8402143 ,  0.61268285, -0.66597616, -0.38463943,
        0.0787466 , -0.02663483, -0.05805221, -0.04350748,  0.27732556,
       -0.04046447,  0.13739883,  0.3859958 ,  0.24203808, -0.04912986,
       -0.19599778, -0.15443575,  0.04484465,  0.18371541, -0.0906113 ,
       -0.06202938,  0.34827092,  0.09899499, -0.08396793,  0.36328898,
       -0.17956662,  0.15839435,  0.06376775, -0.27503705,  0.2707607 ,
        0.32002003])
```

The positive and negative values we can see from the above image are the coefficients (results of autocorrelation) in different lags. These are basically the correlation coefficients between the current and previous values.

A positive autocorrelation indicates that an increase in time series is often followed by another increase. If the autocorrelation result is close to 1, it is almost certain to be followed by another increase. In other words, the average time series value is increasing. On the other hand, a decrease is almost undoubtedly followed by another decrease. In other words, the average time series value is decreasing. Part of the "trend" follows trivially.

You can also draw a graph of autocorrelation.

```python
# Autocorrelation graph
fig = plt.figure (figsize = (12,8))
ax1 = fig.add_subplot (211)
fig = sm.graphics.tsa.plot_acf (ts, lags = 40, ax = ax1)
ax2 = fig.add_subplot (212)
fig = sm.graphics.tsa.plot_pacf (ts, lags = 40, ax = ax2)
```

1 or 12 months is the maximum lag, indicating a positive correlation with the 12-month cycle.

There is a clear seasonal pattern and an upward trend. It is not fair enough to claim that the time series is in the stationary pattern. The population ACF has only been defined for a stationary time series model; the ACF sample can be calculated for any time series.

Due to seasonality, autocorrelation at the end of lag 12 is high, and for each multiple of 12, autocorrelation will tend to be high but will continue to decrease.

ARIMA Model Estimation

For Python, some functions automatically determine the ARMA order, but for some reason, it only identifies the "ARMA" order. In other words, the number of times of the difference is not automatically determined.

Well, this data looks like a summation process at a glance, so I will analyze it after taking the difference. The decision on whether to make the difference is discussed again in the last section, "Automatic SARIMA Model Estimation." In summary, you have to do the 'for' loop yourself to find the optimal order.

Here, the difference is determined once with a fixed hit, and then the automatic ARMA order determination function is applied.

```
# Probably the difference process
diff = ts-ts.shift ()
diff = diff.dropna ()

# Execute automatic ARMA estimation function on difference series
resDiff = sm.tsa.arma_order_select_ic (diff, ic = 'aic', trend = 'nc')
resDif
```

The creation of the difference series has already been explained earlier, so it is avoided here. Just note that we are removing the annoying NaN with dropna. You can determine the ARMA order by using the function "arma_order_select_ic."

The arguments are the target data, what to use as the information criterion (otherwise BIC can be used), and the presence or absence of a trend. The original series seemed to have a pattern, but when I took the difference and looked, it seemed that there was no trend, so I did not include it this time.

Now that we know the optimal order, we can rebuild the model.

```
# Model P-3, q = 2 because it is best
from statsmodels.tsa.arima_model import ARIMA
ARIMA_3_1_2 = ARIMA (ts, order = (3, 1, 2)). Fit (dist = False)
ARIMA_3_1_2.paramsm.tsa.arma_order_select_ic (diff, ic = 'aic', trend = 'nc')
resDiff
```

```
const 2.673501
ar.L1.D.#Passengers 0.261992
ar.L2.D.#Passengers 0.367827
ar.L3.D.#Passengers -0.363472
ma.L1.D.#Passengers -0.075005
ma.L2.D.#Passengers -0.924805
dtype: float64
```

However, this model has its drawbacks, as this has not been able to reflect periodic seasonal variations so well, but the autocorrelation of the residuals is clear at a glance.

```
# Check for residuals
# Because it is not SARIMA, the periodicity remains. . .
resid = ARIMA_3_1_2.resid
fig = plt.figure (figsize = (12,8))
ax1 = fig.add_subplot (211)
fig = sm.graphics.tsa.plot_acf (resid.values.squeeze (), lags = 40, ax = ax1)
ax2 = fig.add_subplot (212)
fig = sm.graphics.tsa.plot_pacf (resid, lags = 40, ax = ax2)
```

Remember that the PACF can be used to find the best AR model sequence. The gray shaded horizontal area represents important factors. The vertical spikes express the ACF and PACF values at the relevant time. Only vertical lines that cross the edge of the horizontal gray shade are considered significant.

SARIMA Model Estimation

Before estimating the SARIMA model, there is a caveat. The SARIMA model is calculated using a library called "statsmodels," which we will discuss in the next part of this chapter. However, the SARIMA model is not included unless the version is 0.8.0 or higher.

If you get angry with Python when trying to estimate a SARIMA model, you can upgrade your version of statsmodels. If you are using Anaconda on Windows, launch the command prompt, and execute the following command.

conda install -c taugspurger statsmodels = 0.8.0

Now we are ready to estimate the SARIMA model. For seasonal variation, the order is fixed.

```
# Estimate the SARIMA model by "fixing"
import statsmodels.api as sm

SARIMA_3_1_2_111 = sm.tsa.SARIMAX (ts, order = (3,1,2), seasonal_order = (1,1,1,12)). Fit ()
print (SARIMA_3_1_2_111.summary ())
```

Use the function "SARIMAX." The name has an "X" in it, which means that other external variables can be included in the model, as in regression analysis.

This time we will go with only one variable. Set the order of the ARIMA model with order = (3, 1, 2) and the order of seasonal variation with seasonal_order = (1,1,1,12).

The last "12" in seasonal_order = (1, 1, 1, 12) means "12-month cycle". So, in effect, (sp, sd, sq) = (1, 1, 1).

The result is shown below. During the calculation, I got a warning that the result of the maximum likelihood method did not converge but ignore it for the time being and proceed.

```
                           Statespace Model Results
==============================================================================
Dep. Variable:                    Passengers   No. Observations:          144
Model:             SARIMAX(3, 1, 2)x(1, 1, 1, 12)   Log Likelihood       -502.990
Date:                        Sat, 27 May 2017   AIC                      1021.980
Time:                                18:50:26   BIC                      1045.738
Sample:                            01-01-1949   HQIC                     1031.634
                                 - 12-01-1960
Covariance Type:                          opg
==============================================================================
                 coef    std err          z      P>|z|      [0.025      0.975]
------------------------------------------------------------------------------
ar.L1          0.5383      1.665      0.323      0.747      -2.725       3.802
ar.L2          0.2830      0.969      0.292      0.770      -1.617       2.183
ar.L3         -0.0346      0.407     -0.085      0.932      -0.832       0.762
ma.L1         -0.9231      1.690     -0.546      0.585      -4.236       2.390
ma.L2         -0.0519      1.658     -0.031      0.975      -3.302       3.198
ar.S.L12      -0.8773      0.285     -3.077      0.002      -1.436      -0.318
ma.S.L12       0.7839      0.370      2.116      0.034       0.058       1.510
sigma2       124.6735     14.315      8.709      0.000      96.616     152.731
==============================================================================
Ljung-Box (Q):                       51.03   Jarque-Bera (JB):          13.51
Prob(Q):                              0.11   Prob(JB):                   0.00
Heteroskedasticity (H):               2.60   Skew:                       0.14
Prob(H) (two-sided):                  0.00   Kurtosis:                   4.55
==============================================================================
```

The first thing that stands out is the header called "Statespace Model (SSM) Results," and it seems that you and the time before the statespace model came out, but it seems that you are using a Kalman filter internally for parameter estimation.

By using the statespace model, the Box-Jenkins method can be handled uniformly. You probably use this for actual calculations. The summary results show a list of information criteria, such as AIC, as well as estimated coefficients. If it is said to be a perfect result, unfortunately, it does not, but we will proceed as it is.

The autocorrelation of the residuals should be almost delicate.

```
# Check for residuals
residSARIMA = SARIMA_3_1_2_111.resid
fig = plt.figure (figsize = (12,8))
ax1 = fig.add_subplot (211)
fig = sm.graphics.tsa.plot_acf (residSARIMA.values.squeeze (), lags = 40, ax = ax1)
ax2 = fig.add_subplot (212)
fig = sm.graphics.tsa.plot_pacf (residSARIMA, lags = 40, ax = ax2)
```

Now we have a model, so let's make a prediction.

```
# Forecast
pred = SARIMA_3_1_2_111.predict ('1960-01-01', '1961-12-01')

# Illustrate actual data and forecast results
plt.plot (ts)
plt.plot (pred, "r")
```

We use the "predict (forecast start date, forecast end date)"
function for the forecast as it is, but note that the forecast target date
must start from the "date included in the original sample." Data is

available through December 1960.

Dynamics = the false argument causes the predictions to be made in one step, i.e., the predictions are made at each point using the complete history up to that point.

However, it is possible to obtain a reasonable representation of our predictive power by using dynamic predictions. In this case, we only use time series information up to a certain point, and the predictions are then made using values from previous prediction data. Let's try the dynamic prediction calculation.

Determination of SARIMA Model Order by Brute Force Method

Finally, let's write our program to determine the order of the SARIMA model; all we must do is change the ordering by brute force, calculate the Akaike information criterion (AIC), and compare. But there are a few points that we can add, so I hope we can share them.

The Akaike information criterion is a means of the goodness of fit of a statistical model. It can indeed be pointed out that it defines the relationship between bias and variance within the structure of the model, or with regards to the accuracy and complexity of the model.

The AIC is not a model analysis in the perception of hypothesis testing. Instead, it generates a method for the divergence between the models of a tool for model options. Provided a set of facts, different options of models could be categorized depending on their AIC by using the model that has the minimal AIC, which is the best.

First, prepare a container for storing the AIC. If you decide the maximum order, you can calculate and calculate how many models of the model will be created. Therefore, calculate the number of repetitions in advance and secure the number of lines by that amount.

The following rules describe the degree:

ARIMA (p, d, q)

Season (sp, sd, sq)

Where:

- p is the degree of the autoregressive model: AR (p),
- q is the degree of the moving average model: MA (q), and
- d is the number of times to make the difference: I (d).

```
# Find SARIMA order that minimizes AIC by brute force
max_p = 3
max_q = 3
max_d = 1
max_sp = 1
max_sq = 1
max_sd = 1

pattern = max_p * (max_q + 1) * (max_d + 1) * (max_sp + 1) * (max_sq + 1) * (max_sd + 1)

modelSelection = pd.DataFrame (index = range (pattern), columns = ["model", "aic"])
pattern
```

When you do this, the number of patterns will be "192". That is, the SARIMA model is estimated 192 times repeatedly. Setting the maximum order to "3" is just a rule of thumb. Just keep in mind that increasing this takes more time.

Regarding the degree of the season, we have a small image because there is an image that does not increase. If you are confident in your computer resources, set a higher number. In the formula for calculating the number of patterns (line 9), max_p is not incremented by one, but this is not a typo.

When the AR term of the ARIMA model was executed with the order set to 0, an error occurred frequently. Therefore, the AR term was looped on the condition that the minimum order was set to 1, so the calculation can be performed only for such slightly distorted times.

Here is the code to actually estimate the SARIMA model, just turn the loop.

```
# Automatic SARIMA selection
num = 0

for p in range (1, max_p + 1):
    for d in range (0, max_d + 1):
        for q in range (0, max_q + 1):
            for sp in range (0, max_sp + 1):
                for sd in range (0, max_sd + 1):
                    for sq in range (0, max_sq + 1):
                        sarima = sm.tsa.SARIMAX (
                            ts, order = (p, d, q),
                            seasonal_order = (sp, sd, sq, 12),
                            enforce_stationarity = False,
                            enforce_invertibility = False
                        ) .fit ()
                        modelSelection.ix [num] ['model'] = "order = (" + str (p) + "," + str (d) + "," + str (q) + "), season = (" + str
(sp) + "," + str (sd) + "," + str (sq) + ")"
```

Python does not use curly braces and expresses loops with indentation, but if you nest so many loops so far, it is a bit hard to read. Review the arguments used in the SARIMAX function. Lines 13 and 14 specify "enforce_stationarity = False" and "enforce_invertibility = False", respectively.

Note that without this, you will get lots of errors and would not be able to turn the loop.

When applying a normal autoregressive model to the summation process, "the model of the stationary process must be estimated as → enforce_stationarity = True" and specify that the model cannot be estimated in the first place. Usually, when dealing with the summation process, the autoregressive model is always applied after taking the difference; therefore, to eliminate the error, this specification can be forcibly specified.

Normally, the result of `` applying the autoregressive model to the integration process " is not selected, but just because it is the smallest AIC does not adopt it as a release, but whether it has a `` decent " order, so check it before using. "Enforce_invertibility" specifies whether to maintain the invertible condition of the moving average model. If this is not set to False, there will be many errors.

In addition, although the error no longer appears, a large number of warnings appear that the parameter estimation result is "not converged." The 15th line of the code, when fitting the model, changing the optimization method as follows + increasing the number of repetitions, reduced the number of warnings. Still, not all, some get warnings.

fit (method = 'bfgs', maxiter = 300, disp = False)

I think that the selected result should not be over trusted but must be analyzed with some doubt. Results come out in minutes. If you pick up the smallest AIC model, it will look like this.

```
# AIC smallest model
modelSelection [modelSelection.aic == min (modelSelection.aic)]
```

model aic

187 order = (3,1,3), season = (0,1,1) 898.105

order = (3,1,3), season = (0,1,1) became the AIC minimum model, so rebuild the SARIMA model again using this order.

```
#
#
bestSARIMA = sm.tsa.SARIMAX(ts, order=(3,1,3), seasonal_order=(0,1,1,12), enforce_stationarity = False, enforce_invertibility = False).fit()

print(bestSARIMA.summary())
```

Here is the result.

```
                          Statespace Model Results
==============================================================================
Dep. Variable:                      Passengers   No. Observations:          144
Model:           SARIMAX(3, 1, 3)x(0, 1, 1, 12)  Log Likelihood        -441.052
Date:                      Sat, 27 May 2017       AIC                    898.105
Time:                              18:53:24       BIC                    921.863
Sample:                          01-01-1949       HQIC                   907.759
                               - 12-01-1960
Covariance Type:                        opg
==============================================================================
                 coef    std err          z      P>|z|      [0.025      0.975]
------------------------------------------------------------------------------
ar.L1         -0.2231      0.097     -2.302      0.021      -0.413      -0.033
ar.L2         -0.1642      0.108     -1.515      0.130      -0.377       0.048
ar.L3          0.7244      0.094      7.704      0.000       0.540       0.909
ma.L1         -0.0837    121.897     -0.001      0.999    -238.997     238.829
ma.L2          0.1221    152.640      0.001      0.999    -299.046     299.290
ma.L3         -0.9797    266.561     -0.004      0.997    -523.429     521.469
ma.S.L12      -0.1583      0.118     -1.337      0.181      -0.390       0.074
sigma2       119.6719   3.26e+04      0.004      0.997    -6.37e+04    6.39e+04
==============================================================================
Ljung-Box (Q):                    36.68   Jarque-Bera (JB):           4.39
Prob(Q):                           0.62   Prob(JB):                   0.11
Heteroskedasticity (H):            1.87   Skew:                       0.16
Prob(H) (two-sided):               0.06   Kurtosis:                   3.90
==============================================================================
```

The residuals do not seem to be a big problem.

```
# Check for residuals
residSARIMA = bestSARIMA.resid
fig = plt.figure (figsize = (12,8))
ax1 = fig.add_subplot (211)
fig = sm.graphics.tsa.plot_acf (residSARIMA, lags = 40, ax = ax1)
ax2 = fig.add_subplot (212)
fig = sm.graphics.tsa.plot_pacf (residSARIMA, lags = 40, ax = ax2)
```

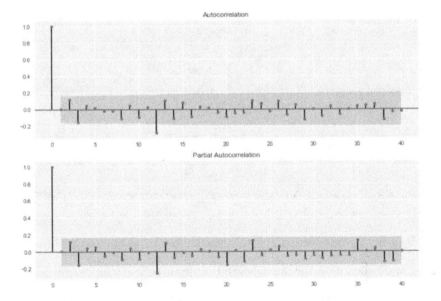

The last is the prediction.

```python
# Forecast
bestPred = bestSARIMA.predict ('1960-01-01', '1961-12-01')
# Illustrate actual data and forecast results
plt.plot (ts)
plt.plot (bestPred, "r")
```

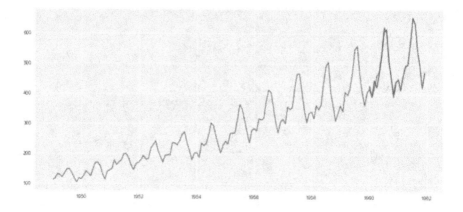

It is important to use a single split when making our training and testing tools because the data MUST guarantee continuity. In fact, the SARIMA model relies on the sequence of the data to make predictions. Looking at the graph above, the model that is ordered as the brute force method does a great job of predicting time series.

Time-Series Prediction With Statsmodels in Python

In this section, we will implement a time-series analysis using the Box-Jenkins method again, and statsmodels, which is also a library of Python. This section will strengthen your skills after getting started with time-series analysis in Python.

A detailed explanation of the method, such as the stationary process and the type of unit root test, will not be explained here because we have already covered that. So, in turn, we will continue just with the implementation.

What is Statsmodels?

It is a library that is a convenient method of implementing a statistical model. It looks like scikit-learn, though scikit-learn is better for machine learning, and the statsmodel is better for lower end statistics.

For this part, I wanted data that had a friendly and easy-to-understand trend and seasonality, so when I Googled sales of ice cream in Kanazawa, Japan (the capital of ice cream of the east), it grabbed my attention.

Library used

```
import statsmodels.api as sm
import pandas as pd
import numpy as np
import requests
import io
from matplotlib import pylab as plt
% matplotlib inline
# Make the graph landscape
from matplotlib.pylab import rcParams
rcParams ['figure.figsize'] = 15, 6
```

The graph should be fixed in a landscape orientation in advance.

Data

```
URL = "https://drive.google.com/uc?id=1MZMKbSQXeVnlAWijTC_hwFCCxPbYNX-Y"

r = requests.get (URL)
row_data = pd.read_csv (io.BytesIO (r.content))
```

The shared link of google drive with the data used is here.

Link:
https://drive.google.com/uc?id=1MZMKbSQXeVnlAWijTC_hwFCCx
PbYNX-Y

```
row_data.head ()
```

Then we get the output as shown below:

	date	earnings	temperature
0	2014/1/1	396.0	3.9
1	2014/2/1	309.0	3.5
2	2014/3/1	447.0	7.4
3	2014/4/1	520.0	12.4
4	2014/5/1	803.0	18.0

Earnings are the sales of ice cream, and the temperature is the average temperature for the month. We will transform this data frame into a form that can be analyzed.

```
# An error occurs when estimating the model unless it is a float type
row_data.earnings = row_data.earnings.astype ('float64')
row_data.temperature = row_data.temperature.astype ('float64')
# make datetime type and index
row_data.date = pd.to_datetime (row_data.date)
data = row_data.set_index ("date")
```

Check the deformed data frame.

```
data.head ()
```

date	earnings	temperature
2014-01-01	396.0	3.9
2014-02-01	309.0	3.5
2014-03-01	447.0	7.4
2014-04-01	520.0	12.4
2014-05-01	803.0	18.0

Now let's see the whole picture.

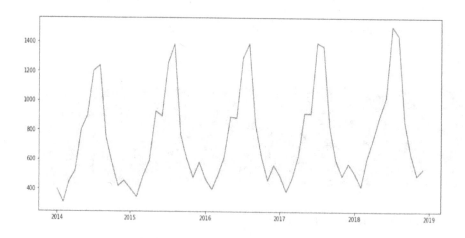

We can see that the data is so easy to understand that its characteristics are clearly visible. Now that the data is ready to process, we will create a model to forecast ice sales. Before that, we will briefly explain some of the important ideas that are necessary for dealing with the time-series data.

Time-series data can be explained by decomposing them into five categories:

1. autocorrelation,
2. periodicity,
3. trend,
4. extrinsic, and
5. white noise.

1. Autocorrelation - The Correlation

The similarity between different points on a time-series is called autocorrelation. Time-series analysis differs significantly from simple regression analysis where it takes into account the relationship factors before and after the data. So, by using autocorrelation, we can express the relationship factors before and after the data is used.

Autocorrelation can be represented by a graph called a correlogram. In statsmodels, you can draw correlograms with graphics plot_acf. But this time, I will not show it because it is similar to autocorrelation, and there is something called "partial autocorrelation." Partial autocorrelation can be plotted with plot_pacf. The autocorrelation of the data this time is shown in the graph below.

```
# Autocorrelation graph
fig = plt.figure (figsize = (12,8))
ax1 = fig.add_subplot (211)
fig = sm.graphics.tsa.plot_acf (data.earnings, lags = 40, ax = ax1)
ax2 = fig.add_subplot (212)
fig = sm.graphics.tsa.plot_pacf (data.earnings, lags = 40, ax = ax2)
```

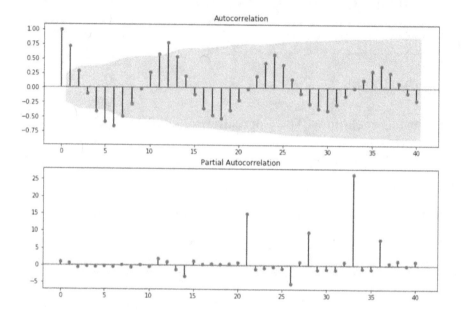

"AutoCorrelation" is the correlogram. By the way, the light blue part indicates the 95% confidence interval. The fact that the data extends outside this confidence interval is statistically significant. The horizontal axis is larger at 12, 24, and 36, indicating that it is similar to the data of 12 months ago. In other words, there is a periodicity every 12 months.

2. Periodicity

Periodicity is also called seasonality, and this refers to a similar change in a 12-month cycle or a periodic change depending on the day of the week. According to the ice cream sales data this time, we can see that there is a seasonality of 12 months cycle.

3. Trends

Trends remain the same, but it is a general trend, in this case, it seems that there is some sort of upward trend.

- ### *Extrinsic*

  1 to 3 were the features in the data, but extrinsic refers to data movement that cannot be explained by autocorrelation alone. For example, it seems that the special feature of ice cream on a popular program has made it possible to sell ice cream.

- ### *White noise*

  The noise that cannot be explained except for the characteristics of the data described above is called white noise, which is also called as residuals.

Using seasonal_decompose of statsmodels, you can quickly decompose time-series data into trend component, periodic component, and residuals. And you can plot as it is!

317

```
# Break down data into trends and seasonal components
seasonal_decompose_res = sm.tsa.seasonal_decompose (data.earnings, freq = 12)
seasonal_decompose_res.plot ()
```

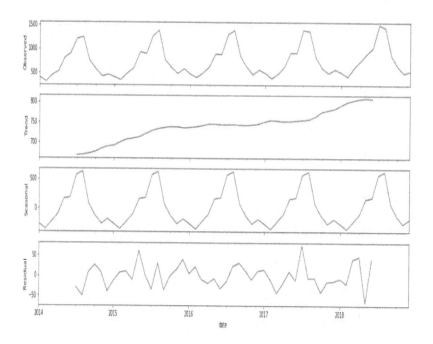

Observed is the original data, and the lower three are the
decomposed components. As expected, there seems to be a rising trend
and a 12-month periodicity.

Trend and seasonal data can be represented by a model called the
"SARIMA" model. Also, this time, the monthly average temperature is
prepared as an extrinsic factor, so the "SARIMAX" model that can take
into account the extrinsic factor will be adopted. By the way, since the

318

average temperature itself should change almost the same as the periodicity of the sales data, it does not make much sense to include it as honest extrinsic.

Make the Data Steady

Although detailed explanations are omitted, data containing trend components are said to be "unsteady" (data that have many zones in one timestep), making it difficult to analyze. On the other hand, "steady" data is easy to analyze, so I want to keep the data in a steady state. Normally, non-stationary data can be converted to the stationary state by taking the difference to remove the trend component. In addition, by performing the "unit root test," we can check whether the data is steady or not.

Below, we are going to perform unit root tests to analyze the properties of stationarity, keeping in mind that when a series or process has a unit root, the series is not stationary, and the least-squares estimators do not have a normal distribution.

A unit root is a stochastic trend in the time series. Sometimes it is called "random drift ride." Therefore, if the series has a unitary root, it presents a systematic pattern that is unpredictable.

So, a time series is stationary if a change in time does not change the shape of the distribution; and unit roots are a cause of non-stationarity.

As the test, let's do a unit root test on the original series without taking the difference. This time, we used the ADF unit root test. The Augmented Dickey-Fuller (ADF) test is a unit root test for a sample of a time series. It is an extended version of the Dickey-Fuller test for a larger and more complex set of time series models.

The augmented Dickey-Fuller (ADF) used in the analysis is a negative number. The more negative the number is, the higher the rejection of the null hypothesis that there is a unit root for a given level of confidence.

```python
# With trend term (up to first order), with constant term
ct = sm.tsa.stattools.adfuller (data.earnings, regression = "ct")
# No trend term, constant term
c = sm.tsa.stattools.adfuller (data.earnings, regression = "c")
# No trend term, no constant term
nc = sm.tsa.stattools.adfuller (data.earnings, regression = "nc")

print ("ct:")
print (ct [1])
print ("-------------------------------------------------")
print ("c:")
print (c [1])
print ("-------------------------------------------------")
print ("nc:")
print (nc [1])
print ("-------------------------------------------------")
```

```
#output
ct:
0.881744298708034
-------------------------------------------------
c:
0.8192957538918915
-------------------------------------------------
nc:
0.9999999896927978
-------------------------------------------------
```

The output value indicates the P-value. In the ADF test, it is a null hypothesis that the data is a unit root process, so if it can be rejected, it can be determined that the data is not a unit root process, which means it is stationary.

The original series is non-stationary because the P-value cannot be rejected in all cases.

Next, shift the data one by one and take the difference, diff () can take the difference of 1 and diff (2) can take the difference of 2 to remove the missing values because they are shifted.

```
diff = data.earnings.diff ()
diff = diff.dropna ()

# With trend term (up to first order), with constant term
ct = sm.tsa.stattools.adfuller (diff, regression = "ct")
# No trend term, constant term
c = sm.tsa.stattools.adfuller (diff, regression = "c")
# No trend term, no constant term
nc = sm.tsa.stattools.adfuller (diff, regression = "nc")

print ("ct:")
print (ct [1])
print (                                                    )
print ("c:")
print (c [1])
print (                              ~                     )
print ("nc:")
print (nc [1])
print (                                                    )
```

```
# Output
ct:
0.005220302847135729

c:
0.0006567788221081102

nc:
0.0019344470286669219
```

All rejected...!

321

By the way, if you plot the difference series, it looks like this.

```
plt.plot (diff)
```

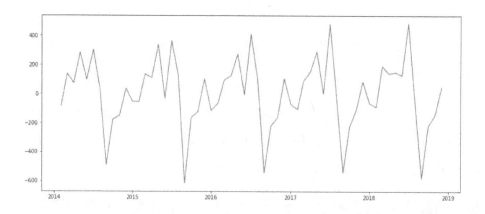

The original series had an upward trend, but the difference series seemed to have a constant average!

Model Estimation

Once the data is steady, all you have to do is estimate the model using SARIMAX in statsmodels. One of the difficulties of the SARIMAX model is that it requires a lot of parameters to set because it can express trends and periodicities. This time, the parameter is estimated semi-automatically using AIC (this is a very common technique).

First, split the data into training and test data.

```
train_data = data [data.index <"2018-06"]
# Test data must include date before test period
test_data = data [data.index> = "2018-05"]
```

Next, estimate the parameters, since the periodicity can be expected to be 12, the other parameters are calculated brute force, and the one with the smallest AIC is used.

```
# Find SARIMA order that minimizes AIC by brute force
max_p = 3
max_q = 3
max_d = 2
max_sp = 1
max_sq = 1
max_sd = 1

pattern = max_p * (max_d + 1) * (max_q + 1) * (max_sp + 1) * (max_sq + 1) * (max_sd + 1)

modelSelection = pd.DataFrame (index = range (pattern), columns = ["model", "aic"])

# Automatic SARIMA selection
num = 0

for p in range (1, max_p + 1):
    for d in range (0, max_d + 1):
        for q in range (0, max_q + 1):
            for sp in range (0, max_sp + 1):
                for sd in range (0, max_sd + 1):
                    for sq in range (0, max_sq + 1):
                        sarima = sm.tsa.SARIMAX (
                            train_data.earnings, order = (p, d, q),
                            seasonal_order = (sp, sd, sq, 12),
                            enforce_stationarity = False,
                            enforce_invertibility = False
                        ) .fit ()
                        modelSelection.ix [num] ["model"] = "order = (" + str (p) + "," + str (d) + "," + str (q) + "), season = (" + str (sp)
+ "," + str (sd) + "," + str (sq) + ")"
                        modelSelection.ix [num] ["aic"] = sarima.aic
                        num = num + 1
```

This can take several minutes.

After the execution, search the data frame for the one with the smaller AIC.

```
modelSelection.sort_values (by = 'aic'). head ()
```

	model	aic
278	order=(3,2,2), season=(1,1,0)	265.434
270	order=(3,2,1), season=(1,1,0)	266.521
286	order=(3,2,3), season=(1,1,0)	268.325
254	order=(3,1,3), season=(1,1,0)	269.788
238	order=(3,1,1), season=(1,1,0)	269.873

Since the model with parameters 3, 2, 2, 1, 1, 0 has the smallest AIC, let's use this to estimate the model.

The estimation of the model can be done with only one line below.

```
SARIMA_3_2_2_110 = sm.tsa.SARIMAX (train_data.earnings, order = (3,2,2), seasonal_order = (1,1,0,12)). Fit ()
```

The model is evaluated using the Ljungbox test, and this test is a null hypothesis, which means that "there is no significant autocorrelation in the residuals" (strictly speaking, it is different...), so I am happy if it is not rejected. Significant autocorrelation of the residuals means that there are features of the data that have not yet been accounted for. You can get the residual with model.resid.

```
# Ljungbox test
ljungbox_result = sm.stats.diagnostic.acorr_ljungbox (SARIMA_3_2_2_110.resid, lags = 10)
df_ljungbox_result = pd.DataFrame ({"p-value": ljungbox_result [1]})
df_ljungbox_result [df_ljungbox_result ["p-value"] <0.05]
```

I want the output of this code to be an empty data frame because I do not want to reject it. If AIC is low, it is not necessarily a good model and not going to help in the analysis.

	p-value
0	0.001934
1	0.005046
2	0.005917
3	0.010949
4	0.018445
5	0.031618
6	0.018867
7	0.000824
8	0.000249
9	0.000466

```
#Model estimation
SARIMA_3_1_3_110 = sm.tsa.SARIMAX (train_data.earnings, order = (3,1,3), seasonal_order = (1,1,0,12)). Fit ()
# Test for residuals
ljungbox_result = sm.stats.diagnostic.acorr_ljungbox (SARIMA_3_1_3_110.resid, lags = 10)
# Result
df_ljungbox_result = pd.DataFrame ({"p-value": ljungbox_result [1]})
df_ljungbox_result [df_ljungbox_result ["p-value"] <0.05]
```

AIC successfully cleared the Ljungbox test with the fourth lowest parameter!

Try to Predict the Sales

Let's use the resulting model to predict sales during the test period.

```
# Forecast
pred = SARIMA_3_1_3_110.predict ('2018-05-01', '2018-12-01')
# Illustrate actual data and predicted results
plt.plot (train_data.earnings, label = "train")
plt.plot (pred, "r", label = "pred")
plt.legend ()
```

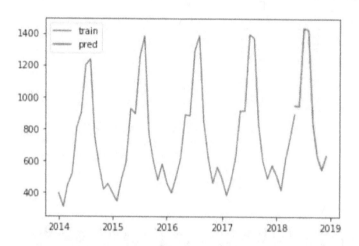

Let's plot together with the correct answer data.

```
# Predictive evaluation
pred = SARIMA_3_1_3_110.predict ('2018-05-01', '2018-12-01')
# Illustrate actual data and predicted results
plt.plot (test_data.earnings, label = "test")
plt.plot (pred, "r", label = "pred")
plt.legend ()
```

As mentioned earlier, the SARIMAX model can take extrinsic factors into account, in this case, only the average temperature, but it may take multiple extrinsic parameters in different analyses.

```
SARIMA_3_1_3_110_with_exog = sm.tsa.SARIMAX (train_data.earnings, train_data.temperature, order = (3,1,3), seasonal_order = (1,1,0,12)). Fit ()
```

By the way, the result is:

```
# Forecast
# Transform data to (7,1)
test_temperature = [[test_data.temperature [i + 1]] for i in range (7)]
pred_with_exog = SARIMA_3_1_3_110_with_exog.predict ('2018-05-01', '2018-12-01', exog = test_temperature)
# Illustrate actual data and predicted results
plt.plot (test_data.earnings, label = "test")
plt.plot (pred_with_exog, "r", label = "pred_with_exog")
plt.legend ()
```

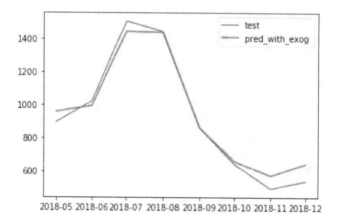

After all, it did not change so much.

I introduced the flow of time-series analysis. Since statsmodels is a really useful library, I hope you can use it to have a wonderful time-series analysis for your purpose.

Time-Series Visualization in Python

Clustering and Converting Time-Series Data with Python

Much of the real-world data is information that changes over time or time-series data. If you look around us, you can easily find data such as behavior data and power consumption data of users who visit your website.

This time, let's examine how to use the "clustering" method to find patterns that repeatedly appear from "time-series data."

Data Used

In recent years, stories like renewable energy and electricity liberalization have been heard. This trend is not only the story of the USA, UK, or Japan, but even the famous Kaggle in the competition of machine learning and data analysis also prefers to use the content related to the power consumption.

329

Link:

https://www.kaggle.com/competitions?sortBy=relevance&group
=all&search=electricity&page=1&pageSize=20

Here, we take advantage of this trend and use the following power consumption data sets among the data sets for machine learning provided by the UC Irvine Machine Learning Repository.

- Data download page URL
    - https://archive.ics.uci.edu/ml/datasets/Electricity LoadDiagrams20112014
- Characteristic
    - Data acquisition period: (2011/1/1, 2015/1/1)
    - No missing measurements
    - If usage starts or ends during the period, the measured value is displayed as 0
- Data form
    - CSV format
    - Column delimiter: semicolon (;)
    - Decimal separator: colon (,) (this is standard in Portugal)

Tools to Use

We use Python 3.6 and Jupyter Notebook as the data analysis environment. The tslearn library is used for clustering time-series data. The necessary tools are listed below, so install them as needed.

- Python (> 3.x) with Jupyter
- Library
  - pandas
  - scipy, numpy, cython (depends on tslearn)
  - tslearn

Acquisition of Data

```
$ wget -nv -O /tmp/LD2011_2014.txt.zip https://archive.ics.uci.edu/ml/machine-learning-databases/00321/LD2011_2014.txt.zip
$ unzip /tmp/LD2011_2014.txt.zip -d /tmp/LD2011_2014
$ ls -l -h /tmp/LD2011_2014
total 679M
-rw-r----- 1 hccho hccho 679M Mar  3  2015 LD2011_2014.txt
drwxrwxr-x 2 hccho hccho 4.0K Mar 17  2015 _MACOSX
```

Looking at the contents of the text, the first column shows time information at 15-minute intervals (no column name), and after that, the power consumption of the client (MT_XXX) is recorded. In the example below, the width is 100 characters, but it is actually longer.

You can see that the data columns are divided by semicolons (';'). The data is reported with one line for each day of the period. The data

331

has missing values; you can note, for example, that for Meter 1 and 2, there is no data.

```
$ head -n 10 /tmp/LD2011_2014/LD2011_2014.txt | cut -c 1-100
"";"MT_001";"MT_002";"MT_003";"MT_004";"MT_005";"MT_006";"MT_007";"MT_008";"MT_009";"MT_010";"MT_011
"2011-01-01 00:15:00";0;0;0;0;0;0;0;0;0;0;0;0;0;0;0;0;0;0;0;0;0;0;0;0;0;0;0;0;0;0;0;0;0;0;0;0;0;0;0;
"2011-01-01 00:30:00";0;0;0;0;0;0;0;0;0;0;0;0;0;0;0;0;0;0;0;0;0;0;0;0;0;0;0;0;0;0;0;0;0;0;0;0;0;0;0;
"2011-01-01 00:45:00";0;0;0;0;0;0;0;0;0;0;0;0;0;0;0;0;0;0;0;0;0;0;0;0;0;0;0;0;0;0;0;0;0;0;0;0;0;0;0;
"2011-01-01 01:00:00";0;0;0;0;0;0;0;0;0;0;0;0;0;0;0;0;0;0;0;0;0;0;0;0;0;0;0;0;0;0;0;0;0;0;0;0;0;0;0;
"2011-01-01 01:15:00";0;0;0;0;0;0;0;0;0;0;0;0;0;0;0;0;0;0;0;0;0;0;0;0;0;0;0;0;0;0;0;0;0;0;0;0;0;0;0;
"2011-01-01 01:30:00";0;0;0;0;0;0;0;0;0;0;0;0;0;0;0;0;0;0;0;0;0;0;0;0;0;0;0;0;0;0;0;0;0;0;0;0;0;0;0;
"2011-01-01 01:45:00";0;0;0;0;0;0;0;0;0;0;0;0;0;0;0;0;0;0;0;0;0;0;0;0;0;0;0;0;0;0;0;0;0;0;0;0;0;0;0;
"2011-01-01 02:00:00";0;0;0;0;0;0;0;0;0;0;0;0;0;0;0;0;0;0;0;0;0;0;0;0;0;0;0;0;0;0;0;0;0;0;0;0;0;0;0;
"2011-01-01 02:15:00";0;0;0;0;0;0;0;0;0;0;0;0;0;0;0;0;0;0;0;0;0;0;0;0;0;0;0;0;0;0;0;0;0;0;0;0;0;0;0;
$ tail -n 10 /tmp/LD2011_2014/LD2011_2014.txt | cut -c 1-100
"2014-12-31 21:45:00";2,53807106598985;22,0483641536273;1,73761946133797;156,50406504065;86,58536585
"2014-12-31 22:00:00";1,26903553299492;22,0483641536273;1,73761946133797;164,634146341463;93,9024390
"2014-12-31 22:15:00";2,53807106598985;22,0483641536273;1,73761946133797;160,569105691057;87,8048780
"2014-12-31 22:30:00";2,53807106598985;22,0483641536273;1,73761946133797;162,60162601626;80,48780487
"2014-12-31 22:45:00";1,26903553299492;22,0483641536273;1,73761946133797;156,50406504065;85,36585365
"2014-12-31 23:00:00";2,53807106598985;22,0483641536273;1,73761946133797;150,406504065041;85,3658536
"2014-12-31 23:15:00";2,53807106598985;21,3371266002845;1,73761946133797;166,666666666667;81,7073170
"2014-12-31 23:30:00";2,53807106598985;20,6258890469417;1,73761946133797;162,60162601626;82,92682926
"2014-12-31 23:45:00";1,26903553299492;21,3371266002845;1,73761946133797;166,666666666667;85,3658536
"2015-01-01 00:00:00";2,53807106598985;19,9146514935989;1,73761946133797;178,861788617886;84,1463414
```

Reading and Checking Raw Data

```python
import logging
logging.basicConfig()
logger = logging.getLogger()
logger.setLevel(logging.WARN)
import itertools
import calendar
from collections import Counter
from datetime import datetime, timedelta
import pandas as pd
import numpy as np
pd.set_option('max_rows', 20)
from tslearn.utils import to_time_series_dataset
from tslearn.clustering import TimeSeriesKMeans
from tslearn.preprocessing import TimeSeriesScalerMeanVariance, TimeSeriesScalerMinMax
import matplotlib.pyplot as plt
import matplotlib.markers as plt_markers
%matplotlib notebook
```

Also, although a bit long, define the following classes for data visualization.

```python
class Visualizer (object):
    def __init__ (self):
        self._markers = itertools.cycle ([
            'o', 'v', '^', '<', '>', '1', '2', '3', '4', '8',
            's', 'p', '*', 'h', 'H', '+', 'x', 'D', 'd', 'P'])
        self._colors = itertools.cycle (['b', 'g', 'r', 'c', 'm', 'y'])
        self._name_to_marker = {}
    def _make_xticks_and_xlabels (self, xs_ticks, xtick_step, xs_labels):
        "" "Generate xtick and xlabel of a specific unit to display x-axis label of graph.
        "" "
        shown_xticks = []
        shown_xlabels = []
        for xtick, xlabel in zip (xs_ticks, xs_labels):
            check_digit = 0
            check_digit |= 1 if xlabel.minute == 0 else 0
            check_digit |= 2 if xlabel.hour == 0 else 0
            check_digit |= 4 if xlabel.day == 1 else 0
            if (xtick_step == 'h' and (check_digit & 1) == 1) \
                or (xtick_step == 'd' and (check_digit & 3) == 3) \
                or (xtick_step == 'm' and (check_digit & 7) == 7):
                shown_xticks.append (xtick)
                shown_xlabels.append (
                    '{} ({})'. format (str (xlabel), calendar.day_abbr [xlabel.weekday ()]))
                logger.info ('Adding xtick, xlabel: {}, {}'. format (xtick, xlabel))
        return shown_xticks, shown_xlabels
    def draw_graphs (self, df, cnt = None, scatter = True,
                xlabel = 'Time', xtick_step = 'm',
                ylabel = 'kW / 15-min', ylog = False,
                figsize = (12,8), dotsize = 3, alpha = 0.3):
        "" "Visualization function for time series data.
        df: Pandas DataFrame. The 'dt' column corresponds to the x axis, and one other column
corresponds to the y axis (one graph).
        cnt: Information to be displayed in addition to the legend name for each y-axis data (eg, the
number of samples belonging to the cluster).
        scatter: Use scatter graph for True and line plot graph for False.
        xlabel: x-axis name.
        xtick_step: h (hour), d (day), m (month).
        ylabel: The y-axis name.
        ylog: If True, set the y axis to log scale.
        figsize: Canvas size.
        dotsize: If scatter = True, set the size of each point.
        alpha: Set the transparency of the graph.
        "" "
        plt.subplots (1, figsize = figsize)
```

```python
        # y-axis: log scale?
        if ylog:
            plt.yscale ('log')
        # Draw a graph.
        xs_labels = df ['dt']
        xs_ticks = np.arange (len (xs_labels))
        ys_columns = df.columns.tolist ()
        ys_columns.remove ('dt')
        for ys_col in ys_columns:
            ys = df [ys_col]
            # Save the graph name and its marker / color information.
            if ys_col not in self._name_to_marker:
                self._name_to_marker [ys_col] = {
                    'marker': next (self._markers),
                    'color': next (self._colors)}
            # If cnt != None, add additional information to the name of each graph.
            if cnt:
                ys_label = '{}: {}'. format (ys_col, cnt [ys_col])
            else:
                ys_label = ys_col
            # Draw a scatter / line graph.
            if scatter:
                plt.scatter (
                    xs_ticks, ys,
                    marker = self._name_to_marker [ys_col] ['marker'],
                    c = self._name_to_marker [ys_col] ['color'],
                    s = dotsize, alpha = alpha,
                    label = ys_label)
            else:
                plt.plot (
                    xs_ticks, ys,
                    marker = self._name_to_marker [ys_col] ['marker'],
                    c = self._name_to_marker [ys_col] ['color'],
                    markersize = dotsize, alpha = alpha,
                    label = ys_label)
            logger.info ('Plotting graph: {}'. format (ys_label))
        # In addition, configure the settings on Canvas.
        shown_xticks, shown_xlabels = self._make_xticks_and_xlabels (
            xs_ticks, xtick_step, xs_labels)
        plt.xticks (shown_xticks, shown_xlabels, rotation = 90)
        plt.xlabel (xlabel)
        plt.ylabel (ylabel)
        plt.grid ()
        lgnd = plt.legend (loc = 'upper left', ncol = 2, markerscale = 3)
        plt.tight_layout ()
        # Show it.
        plt.plot ()
# Create a Visualizer object.
viz = Visualizer ()
```

335

Then, read the above data with Python because it is a CSV file; it can be easily read using Pandas. However, since the delimiter is customized, let's specify the option as follows and load it.

```python
filepath = '/tmp/LD2011_2014/LD2011_2014.txt'
# Load as Pandas DataFrame.
df_raw = pd.read_csv (filepath, sep = ';', decimal = ',')
# Give "str_dt" as the column name of the measurement time column, and give it to Python datetime
# Add a new "dt" column converted to the format.
df_raw = df_raw.rename (columns = {'Unnamed: 0': 'str_dt'})
df_raw ['dt'] = df_raw ['str_dt']. apply (
    lambda s: datetime.strptime (s, '%Y-%m-%d %H:%M:%S'))
# Verify the loading result.
# Try to output only a part because there are many columns.
df_raw [['str_dt', 'dt', 'MT_001', 'MT_002', 'MT_003', 'MT_004', 'MT_005']]
```

	str_dt	dt	MT_001	MT_002	MT_003	MT_004	MT_005
0	2011-01-01 00:15:00	2011-01-01 00:15:00	0.000000	0.000000	0.000000	0.000000	0.000000
1	2011-01-01 00:30:00	2011-01-01 00:30:00	0.000000	0.000000	0.000000	0.000000	0.000000
2	2011-01-01 00:45:00	2011-01-01 00:45:00	0.000000	0.000000	0.000000	0.000000	0.000000
3	2011-01-01 01:00:00	2011-01-01 01:00:00	0.000000	0.000000	0.000000	0.000000	0.000000
4	2011-01-01 01:15:00	2011-01-01 01:15:00	0.000000	0.000000	0.000000	0.000000	0.000000
5	2011-01-01 01:30:00	2011-01-01 01:30:00	0.000000	0.000000	0.000000	0.000000	0.000000
6	2011-01-01 01:45:00	2011-01-01 01:45:00	0.000000	0.000000	0.000000	0.000000	0.000000
7	2011-01-01 02:00:00	2011-01-01 02:00:00	0.000000	0.000000	0.000000	0.000000	0.000000
8	2011-01-01 02:15:00	2011-01-01 02:15:00	0.000000	0.000000	0.000000	0.000000	0.000000
9	2011-01-01 02:30:00	2011-01-01 02:30:00	0.000000	0.000000	0.000000	0.000000	0.000000
...	...	...	...	...	...	...	...
140246	2014-12-31 21:45:00	2014-12-31 21:45:00	2.538071	22.048364	1.737619	156.504065	86.585366
140247	2014-12-31 22:00:00	2014-12-31 22:00:00	1.269036	22.048364	1.737619	164.634146	93.902439
140248	2014-12-31 22:15:00	2014-12-31 22:15:00	2.538071	22.048364	1.737619	160.569106	87.804878
140249	2014-12-31 22:30:00	2014-12-31 22:30:00	2.538071	22.048364	1.737619	162.601626	80.487805
140250	2014-12-31 22:45:00	2014-12-31 22:45:00	1.269036	22.048364	1.737619	156.504065	85.365854
140251	2014-12-31 23:00:00	2014-12-31 23:00:00	2.538071	22.048364	1.737619	150.406504	85.365854
140252	2014-12-31 23:15:00	2014-12-31 23:15:00	2.538071	21.337127	1.737619	166.666667	81.707317
140253	2014-12-31 23:30:00	2014-12-31 23:30:00	2.538071	20.625889	1.737619	162.601626	82.926829
140254	2014-12-31 23:45:00	2014-12-31 23:45:00	1.269036	21.337127	1.737619	166.666667	85.365854
140255	2015-01-01 00:00:00	2015-01-01 00:00:00	2.538071	19.914651	1.737619	178.861789	84.146341

140256 rows × 7 columns

The visualization of these five clients (MT_001 ~ 005) is as follows. In both cases, the data before 2012 is 0, indicating that no contract of consuming electricity was made.

337

```
viz.draw_graphs(
    df_raw[['dt', 'MT_001', 'MT_002', 'MT_003', 'MT_004', 'MT_005']],
    xtick_step='m',
    dotsize=2,
    alpha=0.4)
```

In this case, we will use only the data for the year 2014 (the part surrounded by a brown square) to speed up the experiment.

```
df_sample = df_raw[df_raw['dt'] >= '2014-01-01 00:00:00']
```

Clustering of Time-Series Data and Distance Function

One thing to keep in mind when handling time-series data is the choice of a method to calculate the distance (similarity) between two time-series data. Depending on the task, you may want to recognize the same pattern even if the data has some deviation from the time axis.

For example, in a speech recognition task, the speed of the same utterance varies from speaker to speaker (or even from the same speaker, depending on the situation). A distance function such as Dynamic Time Wrapping (DTW) is used to fill for such a gap.

However, in the power consumption data used this time, deviation from the time axis means an important difference (before and after work or between day and night), so Euclidean distance is used instead of DTW.

Euclidean distance algorithm is the application of a mathematical formula that is applied to the study of graphs and is extremely useful for discovering the similarity between two sets of data. Within the functionalities of the Euclidean distance algorithm, we can determine the similarity between two things or pairs of data that stand out. With it, we can get data schemas that identify elements that have similar characteristics.

Before clustering, convert the power consumption data of each client to a one-dimensional vector representation. This is the input data of the tslearn library used this time. Also, if all data values are 0, they are excluded as exceptions.

The following code will convert the data of each individual client into one dimensional vector representation.

```python
def get_target_column_names (cols):
    columns = cols.tolist ()
    columns.remove ('str_dt')
    columns.remove ('dt')
    return columns
def remove_all_zero_ys (yss):
    filtered_yss = []
    for ys in yss:
        all_zero = True
        for y in ys:
            if y [0]! = 0.0:
                all_zero = False
                break
        if not all_zero:
            filtered_yss.append (ys)
    return np.array (filtered_yss)
# X-axis data.
xs = df_sample ['dt']
# Y-axis data (multiple).
y_columns = get_target_column_names (df_sample.columns)
yss = to_time_series_dataset (
    df_sample.as_matrix (y_columns) .T)
# Delete all zero data
filtered_yss = remove_all_zero_ys (yss)
```

Next, cluster the data with a large number of classes as 12 (this time, we did not find the optimal number of clusters, but we actually need to do this by trial and error). At this time, fix the random value used when initializing the cluster center for the reproducibility of the experiment.

A random seed determines the starting point when a computer generates a series of random numbers. It can be a random number, but it usually comes from the seconds of a computer system's clock.

Random numbers or data generated by the Python random module are not really random, but pseudo-random (i.e., PRNG), i.e., deterministic. It generates numbers from a certain value. This value is nothing more than a starting value (seed value), i.e., the random module uses the starting value as a basis to generate a random number.

In general, the seed value is the previous number produced by the generator. However, when using the random number generator for the first time, there is no previous value. Therefore, the current system time is used as the starting value.

```
n_clusters        = 12
clustering_metric = 'euclidean'
rand_seed         = 12345
km = TimeSeriesKMeans(
      n_clusters=n_clusters,
      metric=clustering_metric,
      random_state=rand_seed)
km.fit(filtered_yss)
```

Let's visualize the clustered result.

```python
# Create a DataFrame with columns of `dt`, `cluster-0`, `cluster-1`, ...
data = []
for cluster_x in km.cluster_centers_:
    data.append (
        [point_x [0] for point_x in cluster_x]
    )
columns = ['cluster-{}'. format (idx) for idx in range (n_clusters)]
clusters = np.array (data) .T
df_clusters = pd.DataFrame (clusters, columns = columns)
df_clusters ['dt'] = xs.values
# Calculate the number of data belonging to the cluster.
cnt = Counter (km.labels_)
cluster_labels = {}
for k in cnt:
    cluster_labels ['cluster-{}'. format (k)] = cnt [k]
# Display graph.
viz.draw_graphs (df_clusters, cnt = cluster_labels,
            xtick_step = 'm',
            alpha = 0.4)
```

The following graph is an enlarged version of the data for the first 30 days in the above graph. The following characteristics appear from these two graphs.

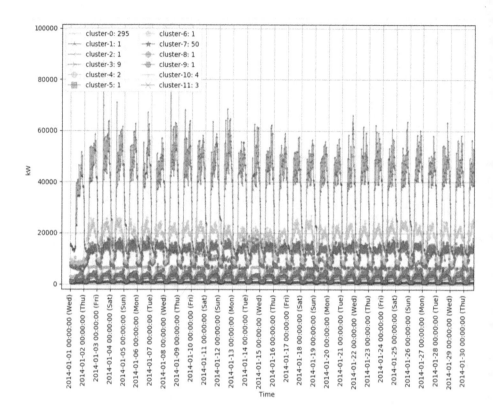

1. The absolute value is very high, and the cluster that peaks in summer (cluster-1) is prominent.

2. There are only half clusters containing a plurality of time-series data. Also, looking at the second graph, in many cases (cluster-1, 2, 5, 6) in which one cluster was generated from one data were much higher in absolute value (10,000 ~ 70,000kW or more).

Clients belonging to cluster-1 require up to 40,000 kW of power consumption in summer, which is about 13,000 times the maximum power consumption of ordinary households (100 V x 30 A) of 3 kW. Considering such power consumption, it is likely that this is a large-scale factory, and in summer, it may need to run coolers to keep the inside temperature normal.

Then, if you plot only cluster-2, 5, and 6 excluding cluster-1, whose absolute value is abnormally high, you will get the following graph.

```
# Plot three large clusters (cluster-2, 5, 6) excluding cluster-1
target_clusters = ['cluster-{}'.format (idx) for idx in [2,5,6]]
filtered_df_clusters = df_clusters [['dt'] + target_clusters]
viz.draw_graphs (filtered_df_clusters, cnt = cluster_labels,
                xtick_step = 'd',
                scatter = False,
                alpha = 0.4)
```

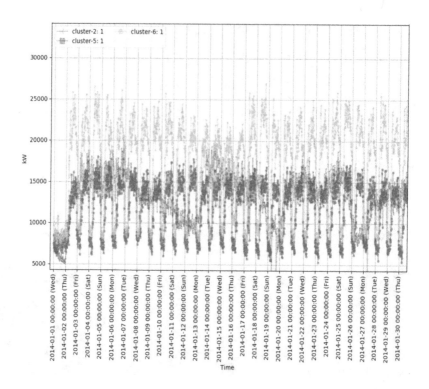

First, you can see that the power consumption is low on January 1 because it is closed. And although cluster-1 and 5 have the difference of the maximum power consumption, it turns out that the pattern is very similar. It can be guessed that they are in the same industry with different scales. However, the power consumption of clients belonging to cluster-2 has a characteristic that falls during the weekend (Saturday and Sunday). It seems that the data is for places that do not open on weekends, such as public institutions.

Next, this time we will look at the characteristics of four clusters (cluster-0, 3, 7, 10) with small absolute values of power consumption. This cluster contains a lot of data. The number of data belonging to the group is on the right side of the cluster name. The first graph is for one year and the second graph is for the first 30 days. The most prominent feature in this graph is that the cluster-0 and 7 show a very stable pattern.

Considering the power consumption, it seems that it is an ordinary household. In addition, cluster-10 has the characteristic of reducing power consumption over the weekend. Although the absolute value is 6-7 times different, it is considered that it is the same industry as cluster-2 seen before.

Let's take a look at the four clusters (cluster-4, 8, 9, 11) where the last power consumption was intermediate. Here I would like to show cluster-8 and 9 and cluster4 and 11 in different graphs. Let's first look at the graphs for cluster-8 and 9. The graphs of cluster-8 and 9 both have large differences between the minimum and maximum values of power consumption and are similar to each other in absolute value.

The fact that it was recognized as another cluster means that the maximum power consumption in February to March in the time-series data belonging to cluster-9 was less than half of the other time, probably recognized as a cluster.

347

However, there is not enough difference in terms of result between cluster 8 and 9 as their trend is very similar, and it indicates that they are the members of the same group perhaps in the commercial sector with approximately the same size.

Link:

https://www.eia.gov/totalenergy/data/monthly/pdf/mer.pc

But what types of industries have such extreme differences in power consumption? Continuing, comparing cluster-4 with 11, since it is very beautifully divided, the patterns tend to be very similar, so, likely, they are also clients of the same industry but of different sizes.

349

In this section, we examined the frequent time-series patterns by applying a clustering method to power consumption data, which is a typical time-series data.

Since the data consisted of only one feature, the power consumption per unit time, I thought at first that it would be easy to get a visible result, but in practice, there were some parts that to you may feel somewhat complex to analyze. But do not worry, just move through the following content of the book, and you will get smoother with practice in time-series analysis with Python.

351

Matplotlib and NetworkX

Now in this section, I will show the use of Python libraries matplotlib (which serves to produce scientific drawings) and networkx (which helps to manipulate and draw complex networks).

As a first example, I will show the code to produce the first figure of this chapter below, which in turn shows a representation of the time-series [2, 3, 10, 2, 4, 5] using matplotlib.

```
import matplotlib . pyplot as plt
y =[2 ,3 ,10 ,2 ,4 ,5]
plt . grid ( True )
plt . plot ( y )
plt . show ()
```

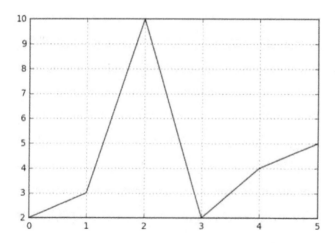

Python has the necessary data structures, such as lists and integers. For example, the second line of the first code example assigns the variable and the list of integers [2,3,10,2,4,5], which in this case, we are interpreting as a time-series. In Python, many libraries extend the capacity of this program. As an example, I can mention the free matplotlib (https://matplotlib.org), which is mentioned in the first line of the first example.

The program has a syntax similar to that of MATLAB™. The next two lines modify a graphic object. With plt.grid(True), a grid is superimposed on the graphic. With plt.plot(y), the lines corresponding to the time-series given by the list we have denoted with y are added. Finally, with plt.show(), the drawing is shown.

Although Python is a somewhat complicated language in terms of complex scientific calculation, it is not necessary to use all its complexity to start using it. I am going to give a slightly more complicated example, which will be useful from now on. To define the visibility graph, it will be convenient to use bars to indicate the time-series values, as shown in the diagram below, the code snippet.

```
import matplotlib.pyplot as plt
y=[2,3,10,2,4,5]
plt.grid(True)
plt.bar(range(len(y)),
        y,
        width=0.2,
        align='center')
plt.show()
```

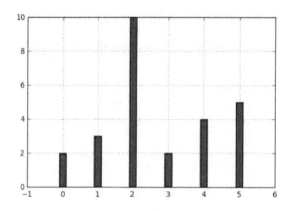

Graphs – Nodes in Python

The next figure shows a drawing of graph P, which we ended in the previous paragraph made with Python and the libraries matplotlib and networkx (the networkx website is located at https://networkx.github.io/). As we can see in the code, shown below, after declaring that we want to use networkx with the import instruction networkx as nx, we can use the functions nx.Graph to create a chart and functions that modify the different elements of a graph (vertices, edges, and labels of the vertices), to create a drawing like before we need to use plt.show() every time.

Note that, in this case, we have created a graph using the nx.Graph function, giving as an argument to the list of edges of the graph to be considered.

```
import matplotlib.pyplot as plt
import networkx as nx
P=nx.Graph([('a','b'),('b','c')])
pos=nx.spectral_layout(P)
nx.draw_networkx_nodes(P,pos,node_size=800)
nx.draw_networkx_edges(P,pos)
nx.draw_networkx_labels(P,pos,font_size=25)
cut = 0.07
plt.xlim(-cut,1+cut)
plt.ylim(-cut,1-cut)
```

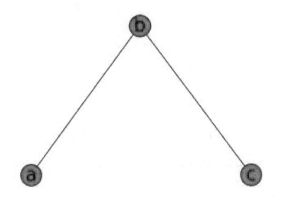

Once the graph has been created in Python, it is possible to use the computer to obtain graph properties. For example:

P. nodes () produces:

['a', 'c', 'b']

P. degree('b') produces:

2

Visibility Graph

In this article (Lacasa and others, 2008), the authors introduce the visibility graph associated with a time-series as a tool for analyzing different properties of the series, using the techniques and terminology of graph theory. On the other hand, they also observe that the visibility graph remains invariant under certain non-essential changes of the time-series, such as translation or rescaling.

Article Link:

https://www.researchgate.net/publication/301232812_From_time_series_to_complex_networks_The_visibility_graph

The visibility graph has its vertices, i.e., the data of the time-series, the vertices can be obtained as V = 0, 1, 2, n 1. The data are declared adjacent to each other in such a way that, in the time-series drawing, the upper parts of their corresponding bars are "visible" to each other, considering the bars as "pairs." Examples are shown in the next two Figures.

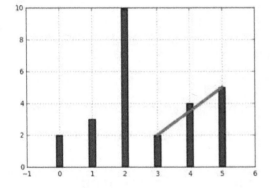

It is then immediate to determine the complete list of the edges of the visibility graph of a time-series, for example, [2, 3, 10, 2, 4, 5], and using networkx to draw a picture of the graph. The drawing is shown in the next figure.

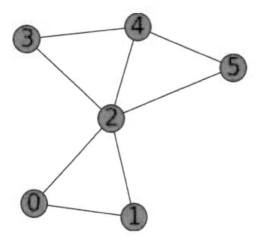

In this case, since the time-series considered has little data, it is feasible to explicitly list the edges of the visibility graph by simply using the time-series drawing. In the following sections, we will show how Python can help us to study more complicated time-series analysis.

The Formal Definition of the Visibility Graph

Determining whether two data are adjacent in the visibility graph in a precise way is an exercise in analytical geometry. For example, (Lacasa and others, 2008) described that the data (t_a, y_a), (t_b, y_b) are declared adjacent in the visibility graph and showed that for all t_c with $t_a < t_c < t_b$ is fulfilled:

$$y_c < y_b + (y_a - y_b)\frac{t_b - t_c}{t_b - t_a}.$$

In Python, we can define a function that determines two different data in a given time-series, and such data are adjacent in the time-series visibility graph. In the code snippet shown below, we define this function. The function is_visible returns true if the data a, b are adjacent and if not false.

```
def is_visible(y,a,b):
    isit = True
    c = a+1
    while isit and c < b:
        isit = y[c]<y[b]+(y[a]-y[b])*((b-c)/float(b-a))
        c = c+1
    return isit
```

On the other hand, the next image shows the code to finish the visibility graph of a time-series, using the function is_visible from the previous code. In this case, a variable called eds. is used to collect the edges in a list. For each data, *a* in the time-series each *b* that is greater than *a* is determined if *b* is visible from *a*, and only it is the edge *(a,b)* added to the eds list. The visibility_graph function finally returns the visibility graph of the time-series ts.

```
def visibility_graph(ts):
    eds = []
    for a in range(len(ts)):
        for b in range(a+1,len(ts)):
            if is_visible(ts,a,b):
                eds.append((a,b))
    return nx.Graph(eds)
```

In the next section, we will apply this code to a much more complicated time-series analysis.

The Logistic Mapping

A simple way to analyze a time-series in a set of real numbers X is using the iterations of a function $f: X \rightarrow X$. To construct a time-series in such a way, an initial condition $x_0 \in X$ is taken, and the subsequent data is defined recursively for $t > 0$ as $x_t = f(x_{t-1})$.

One of the most studied cases, which we include here only as an example, is the logistic mapping, in which $X = [0, 1]$ is taken, that means, the interval of real numbers between 0 and 1, and the function $f(x) = \mu x(1 - x)$, where μ satisfies $0 \leq \mu \leq 4$.

We refer the reader interested in the mapping properties (Devaney (1989), where it is shown that for absolute values of μ, the iterations show the phenomenon of chaos.

Link: https://bu.ac.bd/uploads/BUJ1V4I1/15.%20Hena.pdf

The code below shows how to obtain the iterations of the logistic mapping to produce the next chart, where the time-series obtained with initial condition $x_0 = 0.3$, the value of parameter $\mu = 3.8$, and with 50 iterations is drawn.

360

```
import matplotlib.pyplot as plt
def f(x):
    return 3.8*x*(1-x)
vals=[0.3]
iteraciones=50
for i in range(iterations):
    new = vals[-1]
    vals.append(f(new))
plt.figure(figsize=(10,5))
plt.axis([-1,iterations+1,0,1])
plt.grid(True)
plt.bar(range(len(vals)),
        vals,
        width=0.2,
        align='center')
plt.show()
```

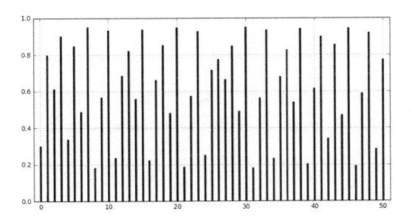

Note that in the code above, in addition to importing the library, a list is created, which contains the data of the time-series. Next, that list can be used to obtain the following figure, which shows the visibility graph of the time-series that originated with the iterations of the logistic mapping. In this case, we note that it is difficult to derive properties of the graph from the drawing alone due to the complexity of the drawing since it is not always clear whether an edge joins two vertices. Therefore, other tools are needed, such as the histogram of degrees, to study a complex graph.

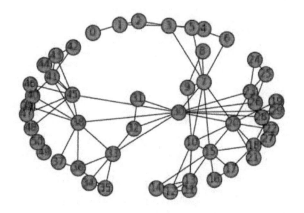

The following code can be used to produce a histogram that is shown in the next image representing the degrees of the previous graph.

In literature, the distribution of the degrees of the vertices is the main tool for studying the graph of visibility, especially in cases where the graph has too many vertices. For example, the situation in which the time-series data are obtained at random from the chaotic case as we consider in this section.

In this part, I showed examples of using the Python program in the study of time-series analysis. There are two possible directions of research regarding the concepts explained here. One is theoretical and refers to studying the properties of the different visibility graphs that have been defined. For example, horizontal visibility graphs are characterized by combinatorial properties.

Another theoretical problem arises when considering that the time-series that can be studied in the computer are limited to having a finite number of terms. However, the concept of the visibility graph naturally extends to the infinite case. It leads to the study of infinite graph properties. For example, an interesting problem is considering how the convergence of a time-series is reflected in the graph.

Data Preprocessing and Time-Series

This part aims to demonstrate practically and understandably how to manipulate a series of historical data to bring out relevant insights. And, finally make a future forecast using machine learning techniques, in particular the ARIMA algorithm (autoregressive integrated moving average). This model is used frequently to make predictions on data where there is seasonality.

The ARIMA algorithm consists of a variation of an ARMA model with the addition of component d for stationarity, this model generates a series of future data using its own past data.

It requires three parameters to be used: p, d, and q.

The first parameter p refers to the number of lags used in the model, i.e., with what periodicity (seasonality) the model has to do. The second parameter d refers to the degree of differentiation, i.e., the number of times the data has had values subtracted in the past. At the same time, q represents the error in the form of a combination of past errors.

For this part, I used Python 3.7 in a Jupyter Notebook configured through the Miniconda package manager.

The Dataset

We start by importing the necessary libraries and datasets, after which the irrelevant columns will be removed from the dataset. Using pandas data, they are transformed into a format called DataFrame, that is, a sort of arrays.

```
# Import of the necessary libraries
import pandas as pd
import numpy as np
import matplotlib. pyplot as plt

plt. style . use ( 'ggplot' )


# Import the data by reading two .csv files via pandas
customers_df = pd. read_csv ( 'customers_export.csv' )
orders_df = pd. DataFrame ( pd. Read_csv ( 'orders_export.csv' ) )
cols_to_drop_cust = [ 0 , 1 , 3 , 4 , 5 , 6 , 7 , 9 , 10 , 12 , 16 , 17 , 18 ]
customers_df. drop ( customers_df. columns [ cols_to_drop_cust ] , axis = 1 , inplace = True )
```

By running these few lines of code and calling the command customers_df.head() we will have access to the first five lines:

	Email	Province Code	Zip	Accepts Marketing	Total Spent	Total Orders
0	@yahoo	BO	26	yes	25.0	1
1	NaN	RO	19	no	0.0	1
2	@gmail	BO	36	yes	25.0	1
3	@gmail.com	BO	28	yes	0.0	0
4	@hotmail	MB	71	yes	25.0	1

One thing that immediately caught my attention was seeing how more than half of the email addresses contained in this dataset have a number that often corresponds to the year of birth. So I decided to write a small algorithm to trace this number and possibly create a graph with the distribution of customers by age.

```
1  # I calculate the age based on the number extracted in the email
addresses
2  cust_age = customers_df. iloc [ :, 0 ] . str . extract ( '(\ d +)' ,
expand = False ) . dropna ( ) . astype ( int ) . tolist ( )
3  for n, i in enumerate ( cust_age ) :
4       if 70 <= i <= 100 :
5           cust_age [ n ] = i + 1900
6  for x in reversed ( cust_age ) :
7       if x> 2010 :
8           cust_age. remove ( x )
9  for x in reversed ( cust_age ) :
10      if x < 70 :
11          cust_age. remove ( x )
12 for x in reversed ( cust_age ) :
13      if x < 1900 :
14          cust_age. remove ( x )
```

After doing a bit of cleaning and turning the list into an object using Pandas.Series I was able to create this chart, including only the first ten bars:

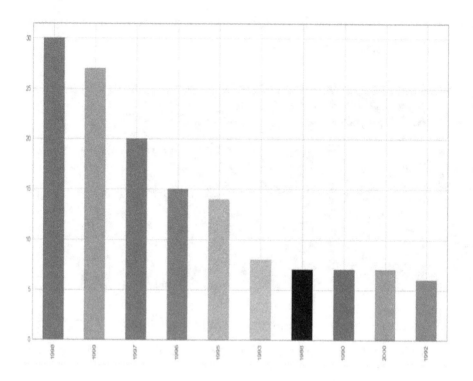

The next step is to analyze which province the customers order from, so just select the column for the region and using the method value_counts():

```
# Customers by region
provinces = pd. DataFrame ( customers_df. Iloc [ :, 1 ] )

# I delete rows containing 'null' data
provinces = provinces. dropna ( axis = 0 , how = 'all' )

provinces = provinces [ 'Province Code' ] . value_counts ( )
```

I get this 'fantastic' pie chart:

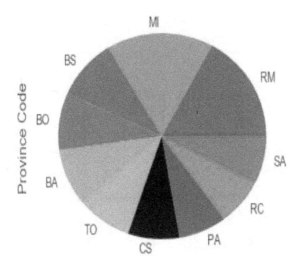

Another interesting point is to find out which are the most sold sizes, this could help to manage the warehouse in a better way, and in the future, it would allow using machine learning techniques to optimize the same.

```
#Convert list to df, drop NaN and plot
tshirt_sizes_df = pd. DataFrame ( tshirt_sizes )
tshirt_sizes_df = tshirt_sizes_df. dropna ( )

plt. plot ( pd. value_counts ( tshirt_sizes_df. values . flatten ( ) ) )
plt. xlabel ( 'Size' )
plt. ylabel ( 'Number of t-shirts' )
plt. title ( 'Number of t-shirts sorted by size' )
```

Number of T-shirt ordered per size

Time-Series Analysis

In this part, I will analyze the dataset based on the order dates, and the dataset contains information from 2016–05–01 to 2018–07–23, therefore more than two years. There are several time-series models and to evaluate which one should be used, it is necessary to use a toolkit that allows you to break down your series into its two components:

- Systematic: components containing consistent and usable trends in our model
- Non-systematic: components of the series that cannot be directly modeled

As we already know that a time-series consists of three systematic components: level, trend, seasonality, and an unsystematic one that is called as the noise.

Before you can use the model though, you need to do some order:

```
# column selection
created_at = pd. read_csv ( r 'C: \ Users \ roberto.sannazzaro \ Desktop \ sales_2016-05-
01_2018-07-23.csv' )

drops = [ 'Sale ID' , 'Order name' , 'Transaction type' , 'Sale type' , 'Sales channel' ,
'Shipping country' , 'Shipping region' , 'Product variant SKU' , 'POS location' , ' Billing
country ' ,
              'Billing region' , 'Billing city' , 'Net quantity' , 'Amount (before discount and
taxes)' , 'Line item discount' , 'Order discount' ,
              'Taxes' , 'Product type' , 'Product vendor' , 'Amount (after discounts before
tax)' , 'Shipping city' ]

created_at. drop ( drops, axis = 1 , inplace = True )
created_at = created_at. groupby ( created_at [ 'Date' ] ) [ 'Amount (after discounts and
taxes)' ] . sum ( ) . reset_index ( )
created_at = created_at. set_index ( 'Date' )
created_at. index = pd. to_datetime ( created_at. index )

y = created_at [ 'Amount (after discounts and taxes)' ] . resample ( 'M' ) . mean ( )

# I create the chart
y. plot ( figsize = ( 15 , 6 ) )
plt. xlabel ( 'Data' )
plt. ylabel ( 'Average sales in EUR' )
```

Projecting the series, it can be seen how the trend is slightly growing, a multiplicative breakdown is right for us in the case of a pattern with low growth/loss, or with a seasonal component can opt for an additive type model.

```
import statsmodels. api as sm
from pylab import rcParams
rcParams [ 'figure.figsize' ] = 20 , 12
decomposition = sm. tsa . seasonal_decompose ( y, model = 'multiplicative' , filt = None ,
two_sided = True , freq = 7 ) # freq = 1
fig = decomposition. plot ( )
plt. show ( )
```

Thanks to this breakdown now, the seasonal component is clearly visible and can be observed as the growing trend. However, it is important to underline that in this case, being aware of the dynamics behind this commercial activity and being mindful of the fact that in the last year there were major changes in marketing and warehouse, I will opt to consider this trend as neutral, as I do not want the model to be influenced by these external factors.

The trend can be positive or negative, linear, or non-linear, so it is very important to know your dataset to understand when enough time has passed to identify a new trend.

At this moment, given the results of the breakdown, we can say that using an ARIMA model makes sense due to the seasonality. For this reason, now it is necessary to choose the parameters P, D, Q (seasonality, trend, noise) for the model.

One of the fastest and most effective systems for estimating these parameters is to carry out a search called "grid search" through different combinations of P, D, and Q values combined with a performance index.

For this, I will use the "Akaike information criterion" (AIC) like before.

The AIC, given a series of models, measures the quality of each in comparison with the other models generated by the combination of p, d, and q. The value of the AIC will allow understanding how each model takes the input data, taking into account the complexity of the model; therefore, the models that will have a better result (fit), using fewer features and that will provide a better (lower) score.

The library for Python pyramid-arima allows you to use this search system quickly, also generates an object with which you can directly call the method fit().

```python
from pyramid.arima import auto_arima
stepwise_model = auto_arima ( y, start_p = 1 , start_q = 1 ,
                              max_p = 3 , max_q = 3 , m = 12 ,
                              start_P = 0 , seasonal = True ,
                              d = 1 , D = 1 , trace = True ,
                              error_action = 'ignore' ,
                              suppress_warnings = True ,
                              stepwise = True )
print ( stepwise_model. aic ( ) )
```

```
t ARIMA: order=(1, 1, 1) seasonal_order=(0, 1, 1, 12); AIC=nan, BIC=nan, Fit time=nan seconds
t ARIMA: order=(0, 1, 0) seasonal_order=(0, 1, 0, 12); AIC=126.466, BIC=127.744, Fit time=0.015 seconds
t ARIMA: order=(1, 1, 0) seasonal_order=(1, 1, 0, 12); AIC=112.105, BIC=114.661, Fit time=0.501 seconds
t ARIMA: order=(0, 1, 1) seasonal_order=(0, 1, 1, 12); AIC=nan, BIC=nan, Fit time=nan seconds
t ARIMA: order=(1, 1, 0) seasonal_order=(0, 1, 0, 12); AIC=121.439, BIC=123.356, Fit time=0.049 seconds
t ARIMA: order=(1, 1, 0) seasonal_order=(2, 1, 0, 12); AIC=nan, BIC=nan, Fit time=nan seconds
t ARIMA: order=(1, 1, 0) seasonal_order=(1, 1, 1, 12); AIC=nan, BIC=nan, Fit time=nan seconds
t ARIMA: order=(1, 1, 0) seasonal_order=(2, 1, 1, 12); AIC=nan, BIC=nan, Fit time=nan seconds
t ARIMA: order=(0, 1, 0) seasonal_order=(1, 1, 0, 12); AIC=124.337, BIC=126.255, Fit time=0.170 seconds
t ARIMA: order=(2, 1, 0) seasonal_order=(1, 1, 0, 12); AIC=117.850, BIC=121.045, Fit time=0.309 seconds
t ARIMA: order=(1, 1, 1) seasonal_order=(1, 1, 0, 12); AIC=117.579, BIC=120.775, Fit time=0.385 seconds
t ARIMA: order=(2, 1, 1) seasonal_order=(1, 1, 0, 12); AIC=119.568, BIC=123.402, Fit time=0.459 seconds
tal fit time: 1.908 seconds

112.10471050293648
```

According to Aike's test estimates, the ideal combination of variables scores 112.10 with a combination of parameters 1, 1, 0, 12.

To verify the effectiveness of the model, it statsmodel provides a method called plot_diagnostics():

It is not perfect, as the line marked with KDE does not precisely follow an average distribution n (0, 1), this can also be seen in the qq graph: the distribution of residues does not differ much, for this reason; the forecasts will have an interval of greater confidence (less precise).

One of the main reasons why the model is not perfect is the lack of data, as the dataset contains information from the exact moment in which the commercial activity started, and therefore in an unstable situation and without any trend identifiable in the first 6 -8 months.

Validation of the Model

To validate the model, and better understand the accuracy of the forecasts, I will compare the forecast sales with the real ones, starting from 2017–12–31:

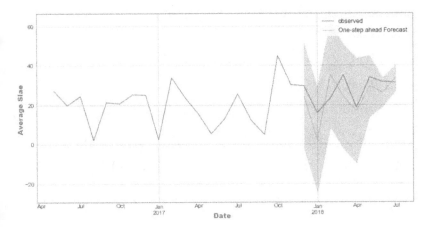

From what is possible to observe, the model predicts real data in the right way and can capture the seasonality presented in the data.

To have a more precise and measurable vision, it is necessary to calculate the R square coefficient, that is the square of the difference between estimated and real values, the more the R square value approaches 1, the better the forecasts will be. However, it is important to specify that a low R square value is not always associated with a low-quality model, as there is no range of standard values in which the R square coefficient must be found, but it depends on each case.

```
y_forecasted = pred. predicted_mean
y_truth = y [ '2017-01-01' : ]
mse = ( ( y_forecasted - y_truth ) ** 2 ) . mean ( )
print ( 'R sqaure: {}' . format ( round ( mse, 2 ) ) )
```

377

I would say not bad, even if it is not perfect.

As we have been able to verify, the model can capture the seasonality. However, the more the forecasts are pushed into the future, the more the confidence interval will increase. In this example, there is only one variable: the average of purchases per week. In contrast, the independent variable is described by the date and seasonality associated with it.

To improve this model and obtain more precise forecasts, a good idea would be, for example, to include an external variable such as weather: this would help to understand if there is a correlation between weather and purchases and therefore, would help (possibly) to allocate the marketing budget more effectively.

Analyzing Time-Series with Python

Throughout this chapter, we will see how to analyze time-series using Python with the knowledge we have acquired from the previous chapters of this book.

Before starting to design a trading strategy, it is necessary to analyze the asset on which we will work carefully. Some questions we can ask ourselves are what is its volatility? What was the minimum level of volatility reached historically? How much did the price go back in the worst session? Or the quintessential question; do you tend to have a trend or anti-trend behavior? By analyzing their past behavior, we answer these questions, and this will allow us to focus on the design of our strategy in a much more efficient way.

Here we are going to create a very simple but also an efficient tool in Python that will allow us to characterize the asset we want to analyze. The ultimate purpose will be to obtain information on whether we should approach the design of the strategy for that asset with a trend or anti-trend approach. We can also find that there is no clearly defined bias, which would mean that there is no distinct advantage to design a system based directly on one of these two approaches.

I have divided the complete analysis of a time-series into the following phases. These will be done following the sequential order shown below.

- **Descriptive Statistics:** In this section, we will visually analyze the distribution of returns and obtain a statistical summary of the series.

- **Volatility analysis:** Here, we will analyze the annualized volatility of the series and its historical evolution. In this section, we will also compare the range in bullish and bearish days to determine if there is asymmetry.

- **Hurst Exponent:** We will apply this tool to measure the persistence of the time-series, which will give us information about the trend, anti-trend, or neutral character of the series.

- **Autocorrelation analysis:** Through this analysis, we will measure the relationship between the last closing price and the previous closing prices. This will let us know what impact the previous closures have on future closures. This analysis also helps us to know if the series has an anti-trend, trend, or neutral behavior.

- **Analysis of the HCLC strategy:** We will program an approach based on a maximum and minimum channel, and we will make a parametric analysis to observe its historical behavior. We will thus check whether there are exploitable patterns in the short, medium, and/or long term using a trend and/or anti-trend approach. Finally, we will register the capital curves of the best trend strategy and the best anti-trend strategy.

- **Summary of results:** We will collect the results obtained in the analysis of the HCLC strategy.

- **Monte Carlo simulation:** Finally, we will make a Monte Carlo simulation of the system that has shown the best behavior (trend or anti-trend).

Used Tools

To design what we have previously called our "tool," we will use a Jupyter Notebook like before. We can access the interactive environment of the Jupyter Notebook through the Anaconda installation. On the official site, there are few versions of Python available for download up to the latest 3.8. Versions below 3 are outdated, so it is not convenient to install one from below 3. Here we are also going to use version 3.7.

We will also access the Yahoo Finance API to download historical end-of-day data. We must consider something important; Yahoo data is not of good quality, and it is usual to find errors in them. We can discover unreal peaks in the quotes, and/or failures in the calculation of splits or dividends. However, here we have used this provider as it also has great advantages; among them, access to the API is free without registration, and it also has a large number of assets with historical lengths. For these reasons, as a starting point, Yahoo seems easy for me to make a quick and convenient exploratory analysis for your convenience.

Descriptive Statistics

Let's start with the creation of the Notebook. The first thing we will do is add the Python libraries that we will use to perform the analysis:

```
import pandas as pd

import numpy as np

import pandas_datareader as dr

import matplotlib.pyplot as plt

# Library to improve chart display

plt.style.use ('seaborn-darkgrid')
```

I will explain shortly again what each of these libraries is for.

- **Numpy:** Acronym of Numerical Python is the library used par excellence for scientific computing and numerical analysis. It includes a large number of functionalities to perform vector and matrix operations.

382

- ***Pandas:*** Built on Numpy, it is a library used to manipulate and analyze data in Python. It offers data structures and operations to manipulate numerical tables and time-series.

- ***Pandas_datareader:*** It is a sub-library that allows obtaining historical data from different online providers and storing us in a DataFrame. From here, you can see the currently supported providers. This library must be installed manually as it does not come by default in Anaconda. If you need it, here you will find a video tutorial to install it.

  Pandas Link: https://pandas-datareader.readthedocs.io/en/latest/remote_data.html
  Video Link: https://www.youtube.com/watch?v=v66K_y3-ezY

- ***Matplotlib:*** It is a library used to generate graphics from lists or arrays in Python and its mathematical extension Numpy.

Next, we will configure the necessary parameters to download the historical data of the asset that we want to analyze. Here we need to configure the name of the asset, and the range of dates that we will download. We will select the SPDR S&P 500 (SPY) ETF from February 1993 to November 2019. As you can see below, through the

Pandas_Datareader library we will download this data and store it in a new DataFrame called "df":

```
# We select the asset to analyze and the dates of study

from datetime import date

Active = 'SPY'

Start Date = '1993-02-01'

End Date = date.today ()

# Download the series and show the Tail.

df = dr.data.get_data_yahoo (Active, start = Start Date, end = End Date)

df = df[~ df.index.duplicated ()] # We remove duplicates in the index (Problems with cryptocurrencies)

df.tail()
```

The df.tail() function will allow us to observe the last five rows of this DataFrame:

	High	Low	Open	Close	Volume	Adj Close
Date						
2019-10-28	303.850006	302.910004	302.940002	303.299988	4.214700e+07	303.299988
2019-10-29	304.230011	302.859985	303.000000	303.209991	4.423800e+07	303.209991
2019-10-30	304.549988	301.989990	303.429993	304.140015	4.958520e+07	304.140015
2019-10-31	304.130005	301.730011	304.130005	303.329987	6.901190e+09	303.329987
2019-11-01	306.190002	304.739990	304.920013	306.140015	7.109800e+07	306.140015

We will now create a new column in our DataFrame that will show the arithmetic percentage return of the adjusted closing price (Adj Close). To do this, we will simply use the Pandas pct_change () function, as shown below:

We create a new column that includes the percentage change of the Closing price under 'R.Arithmetic' column

df ["R. Arithmetic"] = df ["Adj Close"]. pct_change ()

df.tail ()

And ready! Our DataFrame will look like this:

	High	Low	Open	Close	Volume	Adj Close	R. Aritmetico
Date							
2019-10-29	304.230011	302.859985	303.000000	303.209991	44238000.0	303.209991	-0.000297
2019-10-30	304.549988	301.989990	303.429993	304.140015	49585200.0	304.140015	0.003067
2019-10-31	304.130005	301.730011	304.130005	303.329987	69011900.0	303.329987	-0.002663
2019-11-01	306.190002	304.739990	304.920013	306.140015	71098000.0	306.140015	0.009264
2019-11-04	308.000000	307.250000	307.850006	307.494995	29919016.0	307.494995	0.004426

385

Now we have something to start analyzing. We will begin by obtaining a frequency histogram to know the historical distribution that these returns have had. In this histogram, we will also add the probability density function (FDP) of a normal distribution. This will allow us to know if the historical distribution of returns fits, or not, to a normal distribution. Building this graphic through Python is very simple. We will create it by adding the block of code shown below:

```python
import seaborn as sns

from scipy import stats

from scipy.stats import norm



# Delete the first row of the DF; It contains a NaN value in the calculation of the Arithmetic R.

df = df.iloc [1:]



# We draw the frequency chart.

plt.figure (figsize = (15.8))

sns.set (color_codes = True)
```

```
ax = sns.distplot (df ['R. Arithmetic'], bins = 100, kde =
False, fit = stats.norm, color = 'green')
```

```
# We obtain the adjusted parameters of the normal
distribution used by SNS
```

```
(mu, sigma) = stats.norm.fit (df ['R. Arithmetic'])
```

```
# Configure the graphic title, legends, and labels.
```

```
plt.title ('Historical Distribution of Daily Returns', fontsize
= 16)
```

```
plt.ylabel ('Frequency')
```

```
plt.legend (["Normal distribution. fit ($ \ mu = $ {0:.2g}, $
\ sigma = $ {1:.2f})". format (mu, sigma), "R. Arithmetic
Distribution"])
```

As you can see here, we have first imported Seaborn. Seaborn is a Matplotlib-based data-visualization library that provides a high-level interface for drawing attractive and informative statistical graphs. In the following two lines of code, we import the scipy.stats module that contains a large number of statistical functions, and then import norm.

This last function will allow us to create the probability density function (FDP) for the normal distribution. In the following lines of code, we have the necessary instructions to create our frequency histogram (see comments within the code). If we execute this code block, we will obtain a graph like the following:

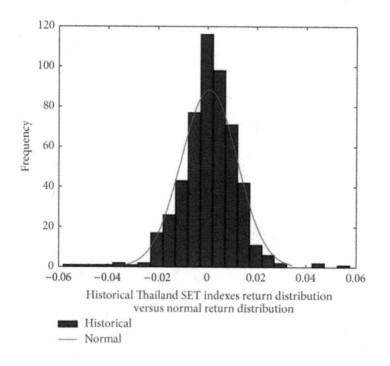

Historical Thailand SET indexes return distribution versus normal return distribution

■ Historical
— Normal

The histogram consists of an x-axis (horizontal) and a y-axis (vertical). In our case, the blue bars of the histogram shows the frequency (y-axis) with which the arithmetic returns are within a certain range of values of the x-axis. The red line shows the FDP of the theoretical normal distribution. This distribution is computed using the average and standard deviation of our historical return series.

Here we can see that our return distribution has some similarity to a normal distribution; however, it does not fit perfectly. In reality, the series of returns that we find in the markets rarely, if not ever, fit perfectly to a normal distribution. In general, they tend to show extreme values that deviate from their average with a higher probability than expected in a normal distribution. This produces distributions with long tails and, therefore, more likely to suffer tail risk. The coefficient of asymmetry and kurtosis help us to know how much our distribution deviates from a theoretical normal distribution. We will get these two values later.

In the analysis of the financial time-series, by simplification, it is based on the assumption that the price series conform to a normal distribution. The main problem that arises with this theory is that it underestimates the risk of obtaining extreme values. That means if we assume that the distribution is normal and its tails are longer than normal, the chances of getting absolute values will be higher than expected. Therefore, when this occurs, it is necessary to rule out normality and use more efficient tools to measure risk. Among which we have the Value at Risk (VaR) or CVaR was applying the historical calculation method.

Next, we will add the following code block to obtain the descriptive statistics of the returns. Here we have added the statistics that I consider most important. However, if we want, we can customize this section by adding those we consider. Inside the code, we will find comments that tell us what the different instructions are for:

```
# We calculate the Composite Annual Growth Rate and the result of Buy and hold.

Years = df ["R. Arithmetic"]. Count () / 252

CAGR = (df ['Adj Close']. Iloc [-1] / df ['Adj Close']. Iloc [0]) ** (1 / (Years - 1)) - 1

print ('> Compound Annual Growth Rate:', '%. 6s'% (100 * CAGR), '%')

print ('> Buy & Hold:', '%.6s'% (100 * (df [' Adj Close ']. iloc [-1] / df [' Adj Close ']. iloc [0])),' % ')


# Calculate the Maximum Drawdown.

Maximum_Previous = df ["Adj Close"]. Cummax ()

drawdowns   =   100   *   ((df   ["Adj   Close"]   -   Maximum_Previous) / Maximum_Previous)

DD = pd.DataFrame ({"Adj Close": df ["Adj Close"],
```

```
                "Previous Peak": Maximum_Previous,

                "Drawdown": drawdowns})

    print ('> Historic MaxDrawdown:', '%.6s'% np.min (DD
['Drawdown']), '%')
```

```
    # We obtain the average, standard deviation, maximum and
minimum value and number of data analyzed:

    print ('> Daily Average:', '%.6s'% (100 * df ["R.
Arithmetic"]. mean ()), '%')

    print ('> Daily Typical Deviation:', '%.6s'% (100 * df ["R.
Arithmetic"]. std (ddof = 1)), '%')

    print ('> Maximum Daily Loss:', '%.6s'% (100 * df ["R.
Arithmetic"]. min ()), '%')

    print ('> Maximum Daily Profit:', '%.6s'% (100 * df ["R.
Arithmetic"]. max ()), '%')

    print ('> Days analyzed:', '%.6s'% df ["R. Arithmetic"].
count ())

    print ('<--------------------------------------------- ----> ')
```

Coefficient of asymmetry and kurtosis of the distribution.

print ('> Asymmetry Coefficient:', '%.6s'% df ["R. Arithmetic"]. skew ())

print ('> Kurtosis:', '%.6s'% df ["R. Arithmetic"]. kurt ())

print ('<--- ----> ')

Theoretical VaR obtained through the normal distribution at 95% and 99% confidence.

print ('> VaR Gaussian Model NC-95%:', '%.6s'% (100 * norm.ppf (0.05, mu, sigma)), '%')

print ('> VaR Gaussian Model NC-99%:', '%.6s'% (100 * norm.ppf (0.01, mu, sigma)), '%')

print ('> VaR Gaussian Model NC-99.7%:', '%.6s'% (100 * norm.ppf (0.003, mu, sigma)), '%')

95% historical VaR and 99% confidence.

print ('> VaR Historic Model NC-95%:', '%.6s'% (100 *
np.percentile (df ["R. Arithmetic"], 5)), '%')

print ('> VaR Historic Model NC-99%:', '%.6s'% (100 *
np.percentile (df ["R. Arithmetic"], 1)), '%')

print ('> VaR Historic Model NC-99.7%:', '%.6s'% (100 *
np.percentile (df ["R. Arithmetic"],. 3)), '%')

If we execute this block of code we will obtain the following results:

```
>Annual Composite Growth Rate: 9.9425%
>Buy & Hold: 1148.6%
>Maximum Historic Drawdown: -55.18%
>Daily Average: 0.0427 X%
>Typical Daily Deviation: 1.1448 X%
>Maximum Daily Loss: -9,844%
>Maximum Daily Benefit: 14,519%
>Days analyzed: 6742
<--------------------------------------------->
> Asymmetry coefficient: 0.0724
> Kurtosis: 11,089
<--------------------------------------------->
>VaR Gaussian Model NC-95%: -1,840%
>VaR Gaussian Model NC-99%: -2.620%
>VaR Gaussian Model NC-99.7%: -3.102%
>VaR Historic Model NC-95%: -1.807%
>VaR Historic Model NC-99%: -3.106%
>VaR Historic Model NC-99.7%: -4.563%
```

In the first section we have the basic statistics of the series; compound annual growth rate, result of the Buy & Hold "strategy", maximum historical Drawdown, etc. Generally, these are the results that we can use to measure the volatility of the stock market for a certain period.

In the lower section, we have the asymmetry coefficient and the kurtosis coefficient. Both serve to know how returns are distributed in our distribution, and how much they deviate from a normal distribution.

The skewness measures the skewness of the distribution of the returns on the average. This value can be positive, negative, or neutral. In a normal distribution, the asymmetry coefficient is 0 since both tails are symmetrically balanced. In our distribution, this value is positive, which indicates that the right tail is longer than the left. In other words, in the right tail, we find values farther from the average than in the left tail, which in this particular case is positive.

Kurtosis (Kurtosis) defines to what extent the different queues distribution queues a normal distribution. In a normal distribution, kurtosis is 3. In our distribution, this value is higher, which indicates that the tails are longer than expected in a normal distribution. In other words, it demonstrates that our distribution contains values that exceed three standard deviations from the average, which are not exceeded with a probability of 99.7% in a normal distribution. When the kurtosis exceeds normal, we can say that the distribution is leptokurtic.

After analyzing both coefficients, we see that the series does not fully adjust to normal. Now let's see what the limitation of estimating the risk of the series assuming that it follows a normal distribution is.

In the last section of the statistics, we have the VaR calculated in two different ways: In the first case, the Gaussian model is used, which assumes that the returns are normally distributed. In the second case, the model based on historical data is used, in which the empirical distribution is used, and in which no assumptions are made about the type of distribution of the series. In both cases, the worst expected loss is obtained with a confidence level of 95%, 99%, and 99.7%.

Here it is observed that with a confidence level of 95%, both methods obtain a very similar value; however, at 99% and 99.7%, the differences are notable. In both cases, the expected result in the empirical model is worse than in the Gaussian model, which confirms again that the left tail of our distribution is longer than normal.

Volatility Analysis

We will now analyze the historical volatility of our time-series. We are primarily interested in knowing the evolution of volatility throughout the period. The code below helps us to create a graph that shows the evolution of the following three values:

- *Annualized volatility:* Standard deviation of the last 14 days multiplied by the square root of 252.
- *The arithmetic mean on annualized volatility (SMA):* Shows the average of the last 126 values of annualized volatility.
- *The closing price adjusted.*

Within the following block of code, we will find different comments that explain the usefulness of the different instructions:

We register Matplotlib converters to adapt the date - DateTime- to the Matplotlib drawing method

from pandas.plotting import register_matplotlib_converters

register_matplotlib_converters ()

We calculate the historical volatility of 14 days, the annualized volatility and its SMA of 252 days.

df ['Historical_ Volatility_14_Days'] = 100 * df ["R. Arithmetic"]. rolling (14).std ()

df ['Volatility_14_Days_Anualized'] = df ['Volatality_Historic_14_Days'] * (252 ** 0.5)

```python
df      ['SMA_126_Volatality_Anualized']      =      df
['Volatality_14_Days_Anualized']. rolling (126).mean ()
```

```python
# We create a graph in which we show the closing price,
annualized volatility and its SMA.

fig, ax1 = plt.subplots (figsize = (15, 8))

ax2 = ax1.twinx ()

ax1.plot (df ['Volatility_14_ Days_Annualized'], 'orange',
linestyle = '-')

ax1.plot (df ['SMA_126_Volatality_Anualized'], 'Green',
linestyle = '-')

ax2.plot (df ['Adj Close'], 'black')
```

```python
# Assign a name for the graph and for the x and y axes.

plt.title ('Historical Evolution of Price and Volatility',
fontsize = 16)

ax1.set_xlabel ('Date')
```

ax1.set_ylabel ('Annualized Volatility', color = 'black')

ax2.set_ylabel ('Closing Price', color = 'black')

Configure the legend, the grid lines, and show the graph.

ax1.legend (loc = 'upper left', frameon = True, borderpad = 1)

ax1.grid (True)

ax2.grid (False)

plt.show ()

This code produces the following graphic:

No doubt, a picture is worth a thousand words. Here we can see the historical evolution of annualized volatility, showing the different peaks and valleys produced throughout history, and their relationship with historical price movements.

The next block of instructions will allow us to analyze this volatility further. Here we will begin obtaining historical volatility. Then we will locate what has been the maximum and minimum value historically achieved and record the dates on which both values have been produced. To complete the analysis, we are also interested in knowing if there is any relationship between volatility and the direction of the session. To answer this question, we will calculate the average percentage range of bullish and bearish days independently. And then, we will build a ratio that will compare both values to know the asymmetry of volatility.

```
# Annualized volatility.

VAM = (252 ** 0.5) * (100 * df ["R. Arithmetic"]. Std ())

print ('> Annualized Volatility:', '%. 6s'% VAM, '%')



# We obtain the minimum and maximum value of annualized volatility as well as the dates

# in which both values are produced.
```

```python
Date_Minimal_Volatility                               =
df.Volatility_14_Days_Annualized
[df.Volatility_14_days_Annualized ==

                df ['Volatility_14_Days_Annualized']. min
()]. index.strftime ('% Y-% m-% d'). tolist ()

Date_Maximum_Volatility                               =
df.Volatility_14_Days_Annualized
[df.Volatility_14_Days_Annualized ==

                df ['Volatility_14_Days_Annualized']. max
()]. index.strftime ('% Y-% m-% d'). tolist ()


print ('> The Minimum Annualized Volatility was', '%.6s'%

    (df ['Volatility_14_Days_Annualized']. min ()),

    '%', 'registered on', Date_Minimum_Voltility [0])

print ('> Maximum Annual Volatility was', '%.6s'%

    (df ['Volatility_14_Days_Annualized']. max ()),

    '%', 'registered on', Date_Maximum_Voltility [0])
```

We get the average over the percentage range of negative days.

```
df ['DaysNegatives'] = np.where (df ['R. Arithmetic'] <0,
100 * (df ['High'] - df ['Low']) / df ['Low'], 0)

df_days_negatives = df.loc [df ['DaysNegatives']! = 0]

DN = df_days_negatives ['DaysNegatives']. Mean ()

print ('> Negative Days Average Range:', '%. 4s'% DN, '%')
```

We obtain the average percentage range of positive days.

```
df ['Positive Days'] = np.where (df ['R. Arithmetic']> 0, 100
* (df ['High'] - df ['Low']) / df ['Low'], 0)

df_days_positives = df.loc [df ['DaysPositives']! = 0]

DP = df_days_positives ['DaysPositives']. Mean ()

print ('> Medium Range Positive days:', '%. 4s'% DP, '%')
```

We calculate the ratio of the range between positive and negative days.

print ('> RDP / RDN Ratio', '%. 4s'% (DN / DP), '%')

After executing this group of instructions, we will obtain the following statistics:

```
> Annualized Volatility: 18,159%
> The Minimum Annualized Volatility was 2.4551% recorded on 2017-08-09
> The Maximum Annual Volatility was 110.78% recorded on 2008-10-28
> Average range of negative days: 1.42%
> Mid Range Positive days: 1.21%
> RDP / RDN ratio 1.17%
```

Here we can see that the annualized volatility of the SPY was 18.16% (using all the history). The minimum volatility was recorded in August 2017, with a value of 2.45%, and the maximum volatility in October 2008 is with a value of 110.78%. We also see that the average percentage range of negative days was 1.42%, and that of positive days of 1.21%. The RDP / RDN Ratio indicates that the average volatility of the negative days was specifically 17% higher than the average volatility of the positive days a normal asymmetry in equity indexes.

Conclusion

The analysis of time-series plays a vital role in the study for forecasting future events in different situations. As we have learned throughout the book, there are several ways or methods to calculate what will be the trend of the behavior of a process under the analysis.

Statistical analysis today is greatly facilitated by tools or software that enable faster processing for further investigation and its extensive graphics capabilities, throughout the book, we showed reports and calculations obtained from different angles and situations.

Analysis of occurrences through time-series for studies of processes of sales, changes in behavior regarding consumption, changes in inflation rates time-series will allow us to analyze in a simple way to forecast the future results and depending on the trend analysis technique we will approach with greater or lesser precision the values that will happen.

Any analysis must also consider that the factors that have been occurring in the period to be evaluated will continue to be influenced in the same way in our future scenario. Any sharp or unexpected change in any of the factors may result in non-compliance with the calculated trends.

When choosing which tools to use for statistical analysis, the most important is the programming language, as this is the core component that will help us to run the analysis flawlessly. Python is characterized by being a dynamic, interpreted, agile, fast-growing language with a massive number of libraries that support the entire ecosystem in the Python world.

Dynamic languages such as Python have the advantage of building MVPs (Minimum Viable Product) in record time. This will allow being at the forefront in the market with a product or service ready in times less than the competition.

To achieve all of the above, we must be able to adapt to the market quickly, continually improving and growing gradually.

Python is completely able to provide us with all the resources we need to get the goals set with the time-series forecasting process.

Although a dynamic tool will indeed facilitate your work, the most important reason when choosing Python as a language is a large number of libraries that are needed to get through your analysis.

We have previously mentioned that Python dynamism plays an important role when choosing this tool over others.

A crucial element when choosing what language to use for our project is the support it has at the data provider level. Python is almost omnipresent as a client to access any service in the cloud where we need to access, either through an API, like any other script or access.

Python has libraries that stand out in the world of Data Science, mathematical and statistical analysis, financial risk analysis, machine learning, Macrodata, etc.

Time-series considered as 'the information superhighway' contains a lot of data that makes the analysis and the processing very complicated. Pandas have handy functions that help you process and obtain essential results for the growth of your statistical projects.

Time Series with Python is amazingly useful and beneficial for every business. Best of luck to you as you explore it.

Resources

https://link.springer.com/chapter/10.1007%2F1-84628-184-9_2

https://petrowiki.org/Production_forecasting_decline_curve_analysis

https://www.machinelearningplus.com/time-series/arima-model-time-series-forecasting-python/

https://www.researchgate.net/publication/301232812_From_time_series_to_complex_networks_The_visibility_graph

https://pdfs.semanticscholar.org/15de/cc2d0f9a1ac16765ed144747ccb97fe1a5d0.pdf

https://www.kaggle.com/competitions?sortBy=relevance&group=all&search=electricity&page=1&pageSize=20

https://archive.ics.uci.edu/ml/datasets/ElectricityLoadDiagrams20112014

https://pandas-datareader.readthedocs.io/en/latest/remote_data.html

https://stackoverflow.com/questions/31413934/ howto-get-fit-parameters-from-seaborn-distplot-fit

https://blog.quantinsti.com/ portfolio-optimization-maximum-return-risk-ratio-python/

www.ingramcontent.com/pod-product-compliance
Lightning Source LLC
LaVergne TN
LVHW050148060326
832904LV00003B/57